SMELL

SMELL
The Tale of a Fading Sense

BJØRN BERGE

Translated by

Lucy Moffatt

REAKTION BOOKS

Published by
REAKTION BOOKS LTD
2–4 Sebastian Street
London EC1V 0HE, UK

www.reaktionbooks.co.uk

First published 2024 as *Lukt*

Lucy Moffatt asserts her moral right to be
identified as the translator of the work

This translation has been published with the financial support of NORLA

NORLA
NORWEGIAN LITERATURE ABROAD

EU GPSR Authorised Representative
Logos Europe, 9 rue Nicolas Poussin, 17000, La Rochelle, France
email: contact@logoseurope.eu

Printed and bound in Great Britain by Bell & Bain, Glasgow

A catalogue record for this book is available from the British Library

ISBN 978 1 83639 250 7

Contents

Preface

To my long-dead grandfather, Einar

One sunny summer's day a few years ago, I was driving through the mountains on the coast of southwest Norway. Finding myself in need of a pee, I turned off into a suitable birch thicket among the smooth rocks.

The instant I opened the car door, it surged towards me: an unmistakable scent of damp nature. Sweetish, spicy and earthy, with a hint of chalk and rusting iron. And all at once I was transported back more than half a century to your cabin by a remote fjord in western Norway. The day was overcast. We were sitting right up close, you with that friendly body of yours, your string vest, shining pate, thick horn-rims, great hairy hands. The little offcuts of wood equipped with paper sails that we sent back and forth across the small pond. Spanish galleons, you insisted in your good-natured growl.

Wild with joy, I jogged into the thicket to find the source of the smell – maybe willow blossoms, bog, fungus, something rotten. But suddenly it was gone, as abruptly as it came. I wouldn't give up, though. I crawled through the blueberry bushes, scraped up handfuls of turf and soil from between the rocks. But all in vain. The smell had evaporated.

Like an addict in the throes of withdrawal, I returned again and again in the days that followed, dawn and dusk, rain and shine. And

I combed the area. But it was no good. The smell could not be summoned up again. It was gone for good. And with it, my childhood summer. It felt as if you'd died a second time.

Introduction

> I collect my tools: sight, smell, touch, taste, hearing, intellect. Night is on its way.
>
> NIKOS KAZANTZAKIS[1]

Pay attention to people when they shake hands and you'll notice that the very next instant, they will touch their nose quickly, as if to inhale a final vestige of the handshake. And that tiny, uncouth gesture repeats itself – with few exceptions – even during the most high-profile encounters, between heads of state or ministers, for example. After handshakes have been exchanged on the red carpet by the steps of the plane, you can almost count down the seconds – rarely more than fifteen – before the hands rise into the air, as discreetly as possible of course.

The smell of the hand communicates.[2] It carries chemical signals with information about mood, stress and, perhaps, disease; things that are otherwise concealed to avoid showing weakness and undermining negotiating positions. The transmission of this knowledge is unconscious and does not result in strategic assessments or shrewd analyses. Yet it *does* inform our gut feeling. And this is often all it takes to choose between friend and foe, war and peace.

Powerful men prefer to present themselves with piercing gazes, gleaming white teeth, elegant ties, wide-shouldered suits and perhaps a Rolex if their country doesn't have much else to offer. Because vision is the sense that rules the world. Or so we think at least.

Soon everything will be image – and nothing but. We live in an age of ocularcentrism, in which the Western human privileges the visual world above all else.[3] It is more important for something to look good than for it to work, whether it's a house, a cake or a person. For the generation currently growing to adulthood, life is like one long catwalk.[4]

The second place in the hierarchy of senses is occupied by hearing, which still has a certain status, whereas taste, touch and smell are ranked lowest – with smell trailing far behind the rest.[5] A global study in 2011 revealed that young people would rather give up their sense of smell than either their computer or their mobile phone.[6]

Our senses and the interplay between them give us direct access to the world and provide a basis for the multitudes of decisions we have to make to exist and survive. Even with the incessant challenge of the media's constant information flow, the only things we can truly rely on when push comes to shove are our immediate sensory experiences. All the rest may well be baloney, bluff and propaganda.

When our senses become imbalanced so that certain of them are suppressed, drowned out or no longer deliver information, we face the risk that the hallucinations will take over. Our brain begins to invent our lives. We are on our way into a fantasy world, according to Finnish architect and philosopher Juhani Pallasmaa.[7]

In an age when artificial intelligence is taking over ever-larger swathes of our lives, even our sense of vision is sometimes forced to capitulate. Yet the sense of smell is in greatest jeopardy. It is in the process of vanishing forever. And it doesn't help that it cannot be digitized.

Humanity has long had an ambivalent relationship to the sense of smell. For long periods, it has been hailed as the finest and most informative of all our senses. At other times, it has been hedged about by taboos and exploited to stigmatize social out-groups such as women, peasants, workers and immigrants.

Its final decline began in the late eighteenth century, during the Age of Enlightenment. Unlike visual impressions, smells proved impossible to organize, categorize and quantify. And it wasn't just that this posed problems for the expanding natural sciences. It was also a question of power, dominance and control – and this has become increasingly true right up until the present day, not just in authoritarian regimes with a particular interest in purity, hygiene and structure, but in more democratic societies.[8]

Smell is associated with the dark, the ungovernable, the erotic, the asocial and the primitive – the very opposite of modernity and progress. In the past few hundred years, great efforts have therefore been made to dilute and combat the many sources of smell, to such an extent that a total absence of smell is now perceived as the ideal. The vestiges that are still tolerated are limited to the functions of cooking, perfuming and deodorizing – always controlled and usually for commercial ends.

This book is a contribution to the battle to liberate our sense of smell in all its beauty and repulsiveness. It opens with an introduction that explains how this sense works, as well as the fundamental conditions that enable smell to exist; it then continues on a voyage of discovery through the history of smells, from the establishment of the first towns in ancient times, via the plague epidemics of the Middle Ages and the Industrial Revolution to the present day.

The book primarily examines European culture. And this is also where the sense of smell has encountered most opposition. Not only have our homes and cities ended up becoming olfactory wastelands, but we are in the process of evolving a kind of sensory illiteracy in this regard. That could prove fateful in a world where established routines and social structures are teetering, and the value of our ability to improvise and unmask disinformation may soon prove more important than ever before.

We must nurture our sense of smell to prevent it from withering away, and it can only be developed through experience.[9] On the plus side, the training process is quick. Whereas it takes several repetitions to consolidate visual impressions, even the smallest sniff is rarely forgotten.[10] Each of this book's twenty chapters introduces a particular smell that has played an important role in most people's lives – a kind of olfactory canon. Each of the smells is presented along with its origin and chemical attributes, and, to stimulate the reader's recognition, the olfactory qualities and character are also described. In addition, the book offers advice about how we can continue to catch a whiff of them.

The Nature of Smell

Out here on the coast of western Norway, storms can blow up out of nowhere even in summertime. When that happens, our sensory apparatus is abruptly roused by the roar of the vast foaming crests that surge up the pebble beaches, the sudden drop in temperature amid clouds of sea spray, the chalk-white scraps of foam that break loose and swirl up into the air like great gulls – but above all, the smell.

It is the smell of the sea. Not of the turbid brackish water between the boats in a marina or of a fjord saturated by salmon farming. This is the primal smell itself, a smell that must have been the same for millions of years – boundlessly fresh and powerful, with clear notes of rotten egg, medical iodine and a whiff of boiled cabbage that is faintly nauseating without being off-putting.

When I stand on the pebble beach below my house, I feel I am being cleansed, from my nostrils all the way to my frontal lobes, and I long to keep having this smell served up to me in boundless quantities to the point of intoxication.

The smell of the ocean originates from the life beneath the surface. The faintly cabbage-like component is created when bacteria feed off the huge populations of plankton that drift upon the world's oceans. That signals a banquet for the entire food chain – from jellyfish to blue whales, from pelicans to turtles. The sharp and slightly metallic smell of iodine stems from the metabolism of algae and marine worms, and when the wind drops and piles of brown kelp lie stranded on the beach quietly fermenting in the spring sunshine, they are accompanied by sharp hints of rotten eggs and overripe cheese. Sometimes, we can also pick up a faintly rancid element of lobsters, crabs, flatfish and other marine creatures. And if the sea is still choppy, we can detect a chalky component stemming from the masses of diatoms that are being churned around in their minuscule silica carapaces.

The Sense of Smell

The nose, that semi-rigid and slightly peculiar growth in the centre of our face, is the only bodily orifice that cannot be closed with either muscles or skinfolds. Instead, it ends abruptly in two arched portals that always stand wide open. When we breathe in – as we do more than 20,000 times a day without giving it a second thought – the airflow passes over the yellowish-brown olfactory mucosa high up in the nasal cavity. It is the size of a postage stamp and is densely packed with slimy sensory cells, which pick up smell substances or odorants. The information is conveyed to the olfactory centre in the brain, which identifies the smell in question, and if it has not been experienced previously, it is stored as a new one. The whole process takes a fraction of a second.

The olfactory centre is directly linked to the limbic system, which instantly processes the impressions to produce emotions and memories. This explains the fresh, spontaneous and individual character of smell. And our sensations can vary depending on our mood. Things may smell different if we are tired or turned on, hungry or well-fed, happy or sad, pregnant or hungover.[11] Sex, age, diet, health and personal experience also play a role. Women are somewhat more sensitive to smell than men, and in all cases, sensitivity peaks after puberty, then declines with age until only vestiges remain by the time we turn eighty.[12]

Twenty per cent of the European population have a deficient sense of smell and around 1 per cent of us suffer from anosmia, a condition in which all smells vanish. This can be caused by growing up in a severely polluted environment, as well as by brain injuries and disease – for example a case of COVID. Some people also experience a distortion of their sense of smell known as parosmia. Suddenly your dinner might smell like garbage, and bananas like rotting meat.

Some smells can also summon up intensely cooling or burning sensations, with a faint tingling of the nostrils. Even people with anosmia can experience this phenomenon, which is not directly linked to the sense of smell, but is associated with irritation of the trigeminal nerves that regulate the sensitivity of our facial muscles. Some odorants such as menthol and ammonia have this dual effect.

Smell and Taste

The sense of smell interacts with the other senses to create overall experiences. The sense of gustation is in a class of its own here. Alone, it is responsible only for a few basic tastes: sweet, sour, bitter, salty and umami. All nuances beyond that – of which there are thousands of course – originate in our sense of smell. For example, a dried apricot tastes of nothing at all if you hold your nose while eating it.

At the same time, conflicts constantly arise between these two closely related senses. Anyone who loves strong cheese, fermented cabbage and Swedish fermented herring knows that. And for many, the smell of parmesan is almost identical to the stink of sweaty socks. If our nose were the only arbiter, such signals could easily become problematic. Only the persistent assurances of our fellow humans persuade us to risk the experience.

Smells

Most of the substances that fill the atmosphere – oxygen, nitrogen, carbon dioxide and water vapour – don't smell of anything at all. In order for a substance to smell, it must first be sufficiently water-soluble to cross the membranes and gain access to the four-hundred-plus sensory cells in the upper part of our nasal cavity. Next, the size and shape must match. Each individual sensory cell is specialized in and receptive to a limited number of odorants. When the odorant matches the cell, it's almost like fitting a key in a lock: the neural circuit is completed and the signal is transmitted to the brain.

A SPOT OF SMELL CHEMISTRY

Every odorant consists of a well-defined molecule that can be described with a chemical formula.[13]

ODORANT	GROSS FORMULA	1. MOLECULAR MASS [U]	2. BOILING POINT [°C]	3. ODOUR DETECTION THRESHOLD [PPB]
Dimethyl sulphide	C_2H_6S	62	37	1–20
Hydrogen sulphide	H_2S	34	60	1–130

The atoms that form the molecules represent irreducible elements. One of the most common elements in odorant molecules is carbon, whose symbol is C, and whose presence makes them organic molecules. In sea air, the organic molecule dimethyl sulphide contributes to the smell of boiled cabbage. In addition to two carbon atoms, it also contains six hydrogen atoms, H, and one sulphur atom, S.

Hydrogen sulphide is an example of an inorganic odorant molecule, and is also a familiar component in the smell of the sea. In this case, the loss of some carbon and hydrogen atoms causes the smell to switch abruptly from rotten cabbage to rotten egg – hardly a drastic change, many would claim. But the addition or subtraction of a single atom can often lead to far greater differences in smells – from violets to oranges, for example, or from apple to vinegar. And to further complicate matters, some molecules with the same chemical formula may smell different, like caraway and mint. In such cases, the shape of the atomic structure is the determining factor.

Odorants are either latent in the mother material or created by external factors. Bacterial attack (decomposition), solar radiation (photochemical reaction) and the addition of oxygen (oxidation) can all transform otherwise odourless molecules into aromatic compounds.

Several conditions must be in place in order for the odorant molecule to escape from the material, disperse into its surroundings and, eventually, be picked up by our olfactory apparatus. Three factors are key: molecular mass, boiling point and odour detection threshold.

1. Molecular Mass

The weight of the molecule, its *molecular mass*, is the fundamental factor that determines whether it can activate the sensory cells in the nose at all. It cannot be higher than 300 u, where *u* is the applicable unit of atomic mass.

Molecular mass also affects how the smell is dispersed in its surroundings. Light molecules spread quickly, whereas heavier molecules are less keen on travelling. However, heavy molecules frequently offer richer and more complex olfactory experiences. They linger for a longer time in homes, clothes, nostril hairs and mucous membranes.

2. Boiling Point

The molecule's ability to escape its mother material is determined by the boiling point of the substance, which ranges from 0°C to 400°C for the most common odorants. The lower the boiling point, the easier it is for the substance to evaporate. At the same time, high atmospheric humidity reduces the boiling point, as does a drop in atmospheric pressure. Odour production is therefore at its most intense in humid, low-pressure weather.

That said, the term boiling point should not be taken too literally. Even where substances have high boiling points, molecules located close to the surface of the material will still evaporate at room temperature, just in lower doses. They will also take longer to fully exit the material.

3. Odour Detection Threshold

Even after gaining access to the sensory cells, the amount of odorant molecules must exceed a given level in order to be perceptible. The required concentration is known as the odour detection threshold and is usually measured in ppb – parts per billion – where 1 ppb corresponds to a billionth of the volume of the air. Still, the detection threshold must be seen as a rough measure since it varies greatly from one individual to another.

But we are rarely dealing with a single odorant. Most olfactory experiences originate from a combination of many such substances. Potatoes emit at least a dozen, roses 250, tomatoes 400 and the smell of coffee, more than 600.[14] And the pattern that all these activated sensory cells form in combination is what creates the overall olfactory impression.

Given the countless combinations that can arise, it is commonly assumed that no two smells are identical. All olfactory impressions have their own subtle singularity, even though the nuances that separate the scent of two lemons will be barely perceptible.[15] Nonetheless, there is little doubt that the human brain can distinguish between a million different smells – some say more, perhaps even an infinite number.

The Presence of Smell

Odorant molecules are often tiny fragments of life – animals, plants and bacteria – which have become dislodged and float freely through the air until they end up in our noses. So when we smell the sea, it is because our bodies have ingested a minuscule portion of the life in the ocean. The same thing happens when we smell another person's body – a thought that can be quite disturbing, since there isn't all that much difference between eating and smelling. In a sense, we are all cannibals.

The sense of smell also tells us about the atmospheric processes that are further transforming the life forms. While solar radiation releases the smell of sulphur from a calm sea, the addition of extra oxygen from a stormy sea generates more iodine – and the more powerful the crashing of the waves, the more intense the smell becomes. In principle, a good nose can replace both eyes and ears. Is the sun shining? Is it day or night? Is a storm raging?

At its most extreme, our sense of smell can also perceive changes in the planet's climate. When evaporation wanes, and increasing acidification and carbon dioxide content alter the living conditions of all the creatures in the sea, the smell also changes.

The Transience of Smell

So you're standing there in the storm, filled to the brim with the powerful smell of the sea – then suddenly it's all over. Sure, the waves are still rolling towards the land, and the roar of the breakers and the cries of the gulls are unchanged, as is the salt taste on your tongue. But the smell has vanished without a trace. At best, you're left with a sensation of freshness, but the smell of the ocean is no longer there. The heavenly gates have closed and the rich aroma is transformed into simple gravitation.

Familiarity

The primary focus of our sense of smell is to detect changes in our surroundings, and we quickly grow accustomed to them. From the moment we enter a landscape, house or car, it is often only a matter of minutes and rarely more than half an hour before the smell vanishes. And that applies even to the most pungent and unpleasant stinks, albeit with variations. Some smells last mere seconds, such as the scents of certain roses, while others, such as camphor and fresh manure, can persist for days. However, the principle remains the same.

The intensity often begins to diminish after the first sniff, or even in the very first fractions of the first sniff, until the brain has located and identified the smell. From an evolutionary standpoint, there's no need for any more.[16] We have the information we require, whether we're talking about sea, roses or cat pee. Or our own smell, for that matter, which we experience only as a whiff of skin and sweat after a session at the gym. Or the smell of our home, which we only notice after we've been out for a hike, or our home town, only after a longish trip. Or our workplace, only for the few minutes after we arrive each morning.

Renewal

And it does no good to strain yourself in the hope of making the smell last or return. Quite the opposite: repeated quick and intense sniffing tends to numb your receptivity even further.

The best way to cleanse your olfactory palate is to go elsewhere – to a house or car that smells different – and then come back again. If you don't have any such options to hand, some simpler procedures may work: if you've walked yourself into a sweat, try a few deep sniffs of your armpit. Taking a few breaths through a woollen scarf will also help. But the best thing of all is to have a special odorant that has as little as possible in common with the olfactory environment you want to explore. Professional 'noses' in the perfume industry often use mild ammonia solutions, camphor or freshly roasted coffee beans.[17]

The Indefinable Nature of Smell

And when you return home after experiencing a stormy sea, how are you supposed to tell people about it? Visual and auditory experiences are no problem: if more precise information is needed, you can describe the colour of the breakers and the size of the waves, or the volume and pitch of the roar. The taste can be described as salty and sour, and the tactile experience of sea spray and flying sand as damp, cold and stinging. But when it comes to smell, your descriptive powers soon fail you. You'll find no terms on the tip of your tongue. Our sense of smell lacks language.

Language

There's no question that a lemon is yellow. But what does it smell like? Like a lemon of course. And coffee smells like coffee. We quickly end up with the phrase *smells like* – like roses, like apples, like freshly laundered sheets, like rotten eggs, like hospitals, like tarmac, like a new car, like a dental surgery, like a wet dog. And all of this is more intimately associated with other sensory experiences – often the sense of vision.

We can also borrow some adjectives from the sense of taste: sour, sweet and bitter. And the smell of fresh chilli pepper can be described as complex, rich and hot, whereas vinegar smells plainer, thinner and colder.

At the same time, some synonyms hint at a kind of quality. While aroma is associated with food and drink, and fragrance is often used to describe perfumes and floral smells, odour tends to be linked to sweat or faeces. If we presume to attach a value to the smell, we can also refer to it as a stink. It all gets pretty vague pretty quickly, and this is roughly as far as we get using everyday language.

Classification

But things looks quite different to a dedicated chemist. By as early as the mid-twentieth century, new analytical techniques, primarily mass spectrometry and gas chromatography, enabled us to identify the structure and properties of many odorants.[18] All were described using chemical formulae and they were often named after the elements they contained. So when a chemist returns home from a stormy sea, he can state with some authority: 'Well, there was plenty of methanethiol in the air today, but perhaps a little less hydrogen sulphide than usual.' This is undoubtedly valuable information, but it hardly succeeds in describing the olfactory experience – it's like describing the flavour of a moist cheesecake as a delightful combination of water and fatty acids.

Ever since the ancient Greeks, philosophers and natural scientists have tried to develop comprehensive classification systems for olfactory experiences, to no avail. More recently, specialists in the wine, perfume and cheese industries have produced various aroma wheels and scent circles, but although they are interesting enough in themselves, they are rarely useful outside these specialist fields. And when the experience cannot be defined and fenced in by language, it cannot be immediately socialized either. The experience of smell remains a primarily personal affair.

Freedom of Smell

Although genetic variations do occur, our sense of smell is less innate than other sensory experiences. No two people will react in exactly the same way, and our responses mostly depend on past experiences, especially in the earliest stages of our lives.[19] Everything our mother eats during pregnancy will influence the scent of the amniotic fluid, and this, together with the smells the newborn child encounters, will shape its odour preferences for life – whether we're talking about garlic, tobacco smoke or shampoo.

Individuality

Our assessment of a smell also depends on the associations it evokes in us, then; and connotations of pain or happiness will determine whether we experience it as nasty or nice. That's why even the dull stench of rotting seaweed may forever be a positive smell for many people because of its associations with summertime and holidays. And if a smell makes us feel nauseous, that doesn't mean it is intrinsically nauseating, but rather that it makes us think of the stomach upset we had after eating something that disagreed with us.[20] If spoiled fish was the culprit, the same nausea will well up the next time someone serves you fish. And it may be a long time before you can stomach the offending foodstuff again.

One evolutionary consequence of this freedom is that we humans rank among the world's most successful generalists. Like rats and cockroaches, we are capable of surviving in pretty much any environment.[21] If we, like the Chinese panda, were only able to eat fresh bamboo shoots, we would soon have died out. Instead, we greet new food smells without any fixed prejudices, yet with a degree of caution.[22] This is neatly exemplified by Scandinavians' long-held scepticism about garlic, which took several decades to overcome.

Culture

With the exception of ammonia and a few other substances, no smells have universally negative connotations; even in those cases, the response comes from our trigeminal nerve rather than our sense of smell. Not even the odour of human excrement is perceived as wholly unpleasant. According to research, European children only begin to react negatively to it between the ages of four and eight, and then only after vigorous lobbying by adults.[23]

Even though personal experience has an absolutely decisive impact on the formation of our odour preferences, culture also comes into play.[24] While global surveys rank the smell of banana as the absolute best, and natural gas as the absolute worst – a ranking I vehemently contest after living solely on bananas for three days during a rough sea crossing in India – major regional differences still apply.[25] Attitudes to the smells of petrol, campfires and cheese vary, and there are more people who dislike than like the smell of honey.[26] In the United Kingdom, the smell of freshly baked bread is a clear winner, with fried bacon in second place and newly mown grass in third.[27]

And the divergences of opinion continue – between population groups, classes, geographical locations and, not least, generations. Britons born in the 1960s and '70s disagree with both their older and younger fellow citizens about the excellence of newly mown grass. Perhaps because they were the generation who had to mow the lawn with those manual monsters that prevailed before the advent of the petrol-driven lawnmower.[28]

The Memory of Smell

Smells are ephemeral and leave no physical traces. But their mental traces are incontestable. Olfactory memories fade much less than other types of memory, and smells can revive moments we have long ago forgotten, often from childhood.[29] They immediately burgeon into full-blown sensory experiences: the shimmer of the afternoon sun on the waves, the keening of the gulls and the grittiness of the warm sand. We feel the presence of the people who were there. Most of all, we relive the feeling of being a small child.

This is often referred to as the Madeleine Effect, named after the famous episode in Marcel Proust's 1913 novel *In Search of Lost Time*.[30] He writes about the scent of the madeleine cake, a little scallop-shaped delicacy, which instantly transports the protagonist back to his childhood visits to Great-Aunt Léonie in Combray: the neat garden, the pond, the pavilion and all the rest. She always served madeleine cakes dipped in lime-flower tea: 'When from a long-distant past nothing subsists, after the people are dead, after the things are broken and scattered, still, alone, more fragile, but with more vitality, more unsubstantial, more persistent, more faithful, the smell and taste of things remain.'[31]

Complete olfactory memories are rarely necessary to spark this. Mere fragments can often be enough, like the faintly rancid smell of seafood and flounder in a fishmongers, or the aroma of bread and cakes in a bakery. Suddenly, you've gone back fifty years in time.[32] The effect is clear and distinct – a phenomenon that is increasingly being used in sense gardens for people with dementia, where the reawakening of good memories helps boost happiness.[33]

The sense of smell's ability to summon up memories is also linked to navigation and sense of direction. Many animals, from migratory birds to wild salmon, use smell for orientation. Every place has its own singular olfactory chemistry, and such navigation is based on an arsenal of accumulated scent memories – almost like programming a GPS.[34] We humans probably have this ability too. And although the processes are not yet fully understood, we have reason to believe that the better you are at recognizing smells, the better your sense of direction.[35]

The Smells that Vanished

Many of our olfactory memories will be shared with others. We can, to a certain extent, agree on them. However, one thing separates us: age.

Most people of my generation – those of us who were born in the 1950s – will probably remember the slightly fishy smell of scorched dust on the electric panel heaters in autumn, the smell of white mothballs, cap guns, newly varnished desks on the first day of school, the terror-inducing smell of the dental surgery, newly smoked sausages in the butcher's shops, mildewy comic books in the outdoor toilet at the holiday cabin, and the smells of fricassee, stuffed cabbage rolls and other old-fashioned dishes once served up on our kitchen table at home. If you were born in the 1960s, you'll remember tinned pineapples, French fries, fish fingers, leather pencil cases, stencil sheets, wax crayons and the dusty scent of newly sharpened pencils, while 1970s children can bond over the profusion of sweet aromas in plastic, glue, hairspray and suntan oil. And all of them will nod in recognition when they catch a waft of the sharp and sweetish smell of a petrol station on a hot summer's day – an experience children born in the 2020s can only dream of.

Some smells have vanished, and others have arrived – like the chemical scent of antibacterial hand gel in supermarkets, the cloying fragrance of e-cigarettes and the sharp metallic smell of EVs. Given rapid changes in technology, dietary habits and materials, smells rarely survive for more than one generation. It feels as if we barely have time to familiarize ourselves with a smell before it's gone. And many of them will never be smelt again. 'I miss Fornebu,' writes Jon Fosse in an ode to the now-defunct airport outside Oslo. 'And the smell of smoke in the café there, I miss that too.'[36]

Marcel Proust was aware of this and, before he settled on the madeleine cake, he mulled several alternatives – including toast and honey, and biscotti, those nutty Italian cookies. Only in the final draft did he opt for the madeleine, which was already a well-established item in the French diet by the end of the nineteenth century.[37] Nowadays, very few people know what he's talking about. But in Henriette Schønberg Erken's cookery book, now more than a century old, we can still find the recipe:

> Take 9 ounces of icing sugar, 10 ounces of flour, 10 ounces of butter, eight eggs and one tablespoon of vanilla sugar. Whip

the egg and icing sugar together. Sift the flour into the melted, cooled butter. Pour the mixture into small moulds (or roll it into balls). Bake for around 6 minutes at 210 degrees Celsius.[38]

And don't forget the lime-flower tea!

1

The Midden: The Smell of Safety

In his autofictional novel *A Death in the Family: My Struggle Book 1*, Karl Ove Knausgård recalls the unique smell of home:

> Even after all that had happened, there were still echoes of the smell I remembered from childhood. As a young boy I had already wondered at the phenomenon: how every house I had been in, all the neighbours' and the family's houses, had a specific smell all of their own which never changed.[1]

Walls of Turf

If you'd found yourself roaming around the turf bogs of Iceland in the late summer 100 or even 1,000 years ago, you'd have found a special experience awaiting you. The Icelandic farm lay half-buried in the ground and, from a distance, you might well have mistaken it for an irregularity in the terrain. Still, you'd know you were on the right track when the spicy scents of heather and bog myrtle borne on the mountain breeze gradually gave way to the sour smoke of a cow-dung-cake fire. And then suddenly there you were on the doorstep, facing a sun-faded trapdoor in the turf wall. And only when the trapdoor opened did the smell become truly intense. 'The stalls beneath were offensive with the damp of earthen walls and refuse fish, while the room upstairs stank of death and a reeking lamp,' according to the midwife who visits the farm of Summerhouses in Halldór Laxness's novel *Independent People*. 'And,' she adds with a shudder, 'down in the stable it was cold and dark, the air sour with the smell of earth, the toadstools flabby.'[2]

In northern Europe, the animal stalls were closely connected to the living quarters until well into the late Middle Ages, and even later in some places. The householders ate their meals among the sheep, dairy cattle and poultry. Here, too, they spun wool, mended nets and read the Bible on Sundays. In his account of a journey through Sweden at the end of the sixteenth century, German merchant Samuel Kiechel spoke of having his face licked by calves, lambs, goats and pigs as he slept.[3] And the 1867 volume of the journal *Fjeld og Hav* (Mountain and Sea) reports that the people of Setesdal, southern Norway, wandered around

> among pigs, rams, chickens, cats, dogs and ducks upon a floor that defies description, whose surface would have sufficed to fill a sizeable field with plentiful manure, and an incredible atmosphere, composed of the thickest and most repulsive odours that can emerge from the human and animal world that this new Noah has collected in his ark.[4]

Back in Iceland, the Scottish geologist Sir George Mackenzie – a man of science so matter-of-fact that he was capable of smelting his mother's diamonds with iron to make steel – speaks of the nauseating conditions he encountered during a stay there in 1810: 'The thick turf walls, the earthen floors kept continually damp and filthy, the personal uncleanliness of the inhabitants, all unite in causing a smell insupportable to a stranger . . . it is wonderful how any thing in the human form can breathe in them.'[5] But the large and poverty-stricken family at Laxness's Summerhouses was used to it. They registered only the sporadic smells of cooking – especially frying fish – of the turf and cow-dung cake that burned alternately in the hearth, and of the occasional pipe-smoking visitor: 'A smell like meadow-sweet, only much more delicious; there is another world in a sweet smell, and the fragrance remained to live and talk when the visitor himself had gone.'[6]

The Home

Whether it's a cave in a mountain or a structure built of turf, timber or bricks, the essential task of the home is to protect us from the threats of the outside world – from heat, cold, rain, wind and thieves. For most people, home represents the safe corner of our lives, the very navel of our existence.[7]

Yet we constantly set out on new adventures and voyages of discovery. And when we come back, our sense of vision already recognizes the facade of the house from a long way off, and our hearing, the creak of the front door. But only our sense of smell can draw us inside it all and give us that absolute and reassuring feeling of having come back home.

The foundations of the smell of home are laid by the smell of the neighbourhood – the smell of the place – from industry, motor traffic, agriculture, forest and sea. It seeps in through cracks, and through windows and doors as they are opened and closed. Once inside the house, it is further acted upon by the smells of the materials used to build and equip the place, both the fresh and the faded: the calciferous concrete, the timber, the bedclothes, the paint, the glue and the plastic materials. The whole medley is rounded off with the smell of the inhabitants themselves and of their activities.

'It is conceivable that it was all to do with habit,' Knausgård suggests,

> using the same soaps, the same detergents, the same perfumes and aftershave lotions, cooking the same food in the same way, coming home from the same job and doing the same things in the afternoons and evenings. If you worked on cars, there would be traces of oil and white spirit, metal and exhaust fumes in the smell, if you collected old books, there would be traces of yellowing paper and old leather in the smell.[8]

The smells of a house vary from room to room, from the introductory smells of damp outerwear and shoes in the hallway through the sharp tang of the wood-burning stove in the living room and the rich food aromas in the kitchen, all the way in to the dry scent of bedclothes in the bedroom. Then it's downstairs to the cellar: 'This was where they had stored the crates of apples, pears, and plums in the autumn,' Knausgård says of his grandparents, 'and combined with the stench of old brick and earth, their exhalations lay like a sub-smell in the house, to which all the others were added and with which they contrasted.'[9]

The Echo of the House

Porous materials in the ceilings, walls, carpets, books and furnishings absorb and temporarily store the smells, like a kind of memory. The capacity is greatest in products made of natural organic materials, such as timber, paper, turf, wool and cotton, but mineral products with rich pore structures, such as plaster, lime and clay, also play an active role.[10] Depending on shifts in atmospheric humidity and temperature, the smells will gradually be fed back into the room, like a long-drawn-out echo of the life of the home. This process is quick for light and volatile substances, whereas heavier and less water-soluble molecules may linger for a long time – like the tars in tobacco smoke, lard from cooking and, above all, the intense acid compounds in human sweat.

In his autobiographical novel *The Notebooks of Malte Laurids Brigge*, the Austro-German writer Rainer Maria Rilke describes the emanations of a demolished house on a rainy Paris street just before the turn of the twentieth century:

> a stubborn, sluggish, musty air which no wind had yet scattered. There the noons lingered, and the illnesses, and the exhalations, and the smoke of many years, and the sweat that trickles down from armpits and makes clothing heavy, and the stale breath of mouths, and the oily smell of sweltering feet. There the pungent odour of urine lingered, and the odour of soot, the grey odour of potatoes, and the heavy, sickening stench of rancid grease. The sweet smell of neglected infants lingered there, the smell of frightened schoolchildren, and the stuffiness from the beds of pubescent boys.[11]

When the internal surfaces of a house are painted with modern vapour-proof paint – often based on acrylic and other plastics – the flow of information is cut off, for a few weeks at least. Sticky organic molecules in the air of the rooms will soon respond by forming a film on top of the material. This, combined with high humidity, will provide ideal conditions for the growth of fungi and microbes, and the already poor smell circulation may become worse than ever.

Bacteria also ensure that the smell of the house changes when new residents move in, no matter how old and aromatic the house might already be. Admittedly, the lingering historical and material components will continue their long-drawn-out processes, but the smell of the newcomers will soon set its seal on the place. Every single one of us carries around our own unique bacterial culture, as unique as a fingerprint. This will not only affect body odour secretions but quickly challenge the other sources of odour in the house. The entire bacterial flora will have been replaced within 24 hours of the new inhabitants moving in.[12] And even an otherwise anonymous hotel room will take on your smell.

Identity

Whether exclusive or brutal, the smell of your home is the one your visitor will always associate with you. If the guest has come from a long way off, it will also be the smell of a different culture, like the stench of cod liver oil and sour earth in the Icelandic peasant's house; the emanations of unwashed wool and goat's milk in a Mongolian yurt; of walls plastered with cow dung among the people of the African savannas; and of the untanned sealskin on the internal walls of Inuit igloos. And if, along the way, the traveller has recognized a smell that reminds them even slightly of their own culture, that will almost certainly have brought them consolation and joy.

Here's what happened some years ago at our architects' office in Lista. We were all men, all full-grown adults, and these facts were reflected in the smell of our only sporadically cleaned toilet – ingrained, cloying, with a sour hint of ammonia. For various reasons, the nursery next door to our office had to borrow our loo for a few days. I can't deny we felt almost ashamed as we listened to the children coming up the stairs in groups, until we suddenly heard one little girl exclaim joyfully: 'Oh! It smells so nice in here – Just like home.'

Midden

Humanity came into being around the campfire. We have used fire for heating and cooking ever since the days of the Neanderthals, across cultures and continents. The open campfire stimulates all our senses, with its flickering flames, its crackle and sputter, its intense heat and, most of all, its smell, that peppery tang of oil and metal. Our blood pressure drops and the conversation grows calmer. Gathering around the fire brings the flock together.

Some years ago, archaeologists conducted a dig at the Stone Age settlement of Vistehola, a cave in a shale knoll a few hundred yards above the shoreline in Jæren, western Norway. It faces south, offering good visibility, and looks like an ideal residence for early Stone Age hunters. With a depth of 8–9 metres (26–30 ft) and space for ten to fifteen occupants, it's like the Stone Age equivalent of a mansion. The archaeologists were initially astonished by the odd acoustics, which require you to whisper to be understood at all. But the really big surprise lay hidden in a heap of refuse – a midden – just inside the opening. In addition to bones, shells, and food and flint waste, they found the remains of a fifteen-year-old boy. He had been placed on the pile along with the rest of the refuse 8,000 years before.[13]

The cave-dwellers must have been keenly aware of the child's process of decay. After a short time, the body's own enzymes kick-started the decomposition of the tissue, giving access to fungi and bacteria. A few days later, swarms of flies and beetles came and laid

DEATH

ODORANT	GROSS FORMULA	MOLECULAR MASS [U]	BOILING POINT [°C]	ODOUR DETECTION THRESHOLD [PPB]
Cadaverine	$C_5H_{14}N_2$	102	179	0.56
Putrescine	$C_4H_{12}N_2$	88	158	0.4

There's little olfactory insight to be gained from a chunk of meat in an airtight jam jar, no matter how thick its rind of fat. Even after a few days, the best you can hope for is a faint, nauseating whiff of the cheesemongers. Things only start happening when you add a few tablespoons of soil, after which you'll soon pick up a hint of what most think of as the smell of death. Created when the bacteria in the soil break down the proteins, fats and carbohydrates in the meat, we perceive it as sharp and intense, yet also thin and cold.

The dominant odorants in the mixture are the nitrogen compounds cadaverine and putrescine, whose names speak for themselves. Cadaverine results from the decomposition of lysine, the amino acid we obtain primarily from food – especially meat, fish and dairy products – whereas putrescine is produced by the decomposition of arginine, the amino acid produced by the body itself.

While putrescine contributes a sharp, ammonia-like note, cadaverine is responsible for an intensely cloying smell – which many would describe as slightly chemical – with a hint of overripe fruit and rotten cabbage.

Although these odorants aren't especially volatile, both have low odour detection thresholds and their stink – as most people would describe it – is penetrating, and detectable from a long way off.

Cadaverine and putrescene can be found in bad breath, too, especially among meat-eaters with a casual approach to oral hygiene. These odorants also contribute to the fishy tang in the vaginal discharge of a woman with a bacterial infection, and are contained, albeit in lower concentrations, in the smell of sperm.

eggs that quickly hatched into myriads of hungry larvae, and soon the flesh decomposed.

The whole process was accompanied by an ever-intensifying, cloying stench that would be appalling by our standards. But we cannot take it for granted that the Stone Age people experienced it that way. They probably did not. They were used to the smell of death. And in their day-to-day life they used food preparation and conservation methods based on putrefaction – for trout, oysters or puffins. A study which shows that European children up to the age of five display no response whatsoever to the smell of putrefaction suggests that this is a culturally acquired aversion.[14]

It may be that burial in the midden, surrounded by his own kin, was seen as the perfect departure from this earthly plane. And perhaps for those left behind, the corpse stench of a brother, mother or grandfather, mingled with fragments of animal fat, leather and wet rock, epitomized the safe embrace of home. Like the smell of the campfire, it brought the flock together, only this time all the way into death.

2

A Feast for the Senses: The Smell of Pleasure

Twenty-nine-year-old archaeologist Rodolfo Lanciani became an overnight sensation in the mid-1870s when he discovered a vast, 2,000-year-old mass grave on the eastern flank of Rome's Esquiline Hill. In open pits, known as *puticuli,* he reported that he had found 'a uniform mass of black, viscid, pestilent, unctuous matter' that had apparently lain exposed to wind and weather. He continued:

> The reader will hardly believe me when I say that men and beasts, bodies and carcasses, and any kind of unmentionable refuse of the town were heaped up in those dens. Fancy what must have been the condition of this hellish district in times of pestilence, when the mouths of the crypts must have been kept wide open the whole day![1]

Cloaca Maxima

Legend has it that Rome was founded in 750 BCE. More than 5,000 years after the Stone Age settlement in Jæren, we have arrived at urban society. The once-scattered collection of tribal peoples has now coalesced into vast, complex and densely packed structures, with order and discipline and division of labour. The cities of ancient Rome are also famed for their thorough cleanliness and hygiene.[2] We are told of an extensive infrastructure of aqueducts that transported crystal-clear water into public baths and fountains before discharging it through vast drainage systems.

With its mighty vault of massive stone blocks, Rome's Cloaca Maxima – the main sewage system – has served as the model for the sanitation systems of Europe's major cities all the way up to the present day.[3] In reality, though, it had little to do with sewage. With the exception of a handful of ornate high-status latrines in the Forum, very few toilets were connected to the system. The primary purpose of the Cloaca Maxima was to prevent the city's low-lying and central districts from flooding during the heavy autumn and winter rains. Apart from serving sporadically as a rubbish dump, it was sometimes used to dispose of the corpses of plague victims during epidemics – and, occasionally, the carcass of some liquidated senator denied honour in death.[4]

Excrement and Waste

In around the year 1 CE, Rome's 450,000 inhabitants produced more than 300 million litres of urine and 59,000 tonnes of excrement per year.[5] This was conveyed in pipes to simple tanks in the back gardens. Because they were also used for food waste, they were built wall to wall with the kitchen, and were constructed in such a way that the fluid could drain freely into the surrounding soil through cracks and openings. They appear to have been functional enough and are reminiscent of modern septic systems involving filtration, but with one crucial difference: they had no water trap. Consequently, although the supply pipe might be flushed out periodically with a bucket of water, the smell will have been intense, especially in the kitchen. The stench must have spread out into the streets, providing the city with a kind of permanent olfactory backdrop.

Outdoors, this will have been amply supplemented by the smell of the almost omnipresent domestic animals: horses, mules, dogs, chickens, ducks and geese. On top of that came the butchers' waste dumps, the piles of fish guts at the fish market, and the stink of all kinds of workshops and small industry, including laundries that used old urine for both cleaning and bleaching. And all the while, countless steaming food wagons rattled back and forth across the city, trailing the stench of garum – the legendarily evil-smelling sauce made from fermented fish guts that was used not just for flavouring but to camouflage spoiled food. While modern fish sauces may well have the same element of umami, the smell of garum was probably in a totally different league.

To cut a long story short, there is little doubt that Rome stank.[6] By our standards at least. Yet the Romans didn't complain, not even when the hot summer winds blew down from the open *puticuli* on the Esquiline Hill. The noise, on the other hand, was a problem. The satirist Juvenal speaks of 'creaking carts in narrow streets. Cursing coachmen in teeming crowds' and claims that 'sleeplessness robs Rome's sick and feeble of their health.'[7] But the smell is never mentioned, other than in one passage where he notes the fresh aroma of newly sawn timber being transported through the streets, and another where he speaks about the minor unpleasantness of sweat in a dense throng.

From the High Middle Ages 1,000 years later until well into the twentieth century, the smells of major European cities changed from one district to the next, almost like the different rooms in a house. They reflected the activities of the particular quarter – some industrial, with glue factories and foundries, others boasting promenades planted with avenues of chestnut. And the neighbourhoods had distinct class affiliations, the boundaries between them clearly discernible to the nose. Roman cities, however, were more homogeneous. With the exception of aristocrats, rich and poor generally lived in the same buildings – the rich on the ground floor and the poor in cramped attic rooms – and workshops, industries and markets were spread evenly throughout the city.[8] So you could expect no major contrasts on a walk through Rome. The city's smell was largely the same everywhere, just as it is in the modern cities of the twenty-first century. Now, though, that isn't because the city's class divisions have been eliminated. Far from it. It's simply because the smells have been banished.

In Rome the only smells liable to cause concern were the ones that emanated from the outside. The architect Vitruvius warned against the morning breezes that wafted over the city at sunrise, describing them as downright unhealthy, when 'they bring with them mists from the marshes and, mingled with the mist, the poisonous breath of the creatures of the marshes.'[9] Meanwhile, burghers who ventured beyond the city limits must be prepared for unpleasant encounters with peasants who stank of goats and heavy garlicky sweat, and they would often carry nosegays of scented flowers to take the edge off the experience.[10]

Red Mist

There was one exception to what was still, after all, a kind of olfactory harmony. The Colosseum and other amphitheatres provided a constant stream of popular entertainment in the form of gladiator battles to the death, often followed by the mass slaughter of leopards, lions, tigers, hippos and elephants. This resulted in a thoroughly distinctive olfactory experience: the smell of newly drained, fresh blood – buckets of it.

The smell of blood often means that something has gone wrong, and it triggers strong feelings in most of us.[11] Take Norwegian army medic Ida Vilhovde, who found herself in the interminable bloodbath of a Norwegian field hospital during the evacuation of Afghanistan in summer 2021: 'A smell permeated the entire hospital, a smell of iron, that clings to you . . . it was absolutely ghastly.'[12]

The organizers at the Colosseum also discovered that the smell of blood was not conducive to restraint and calm. To soothe the spectators, they set up a network of fountains that sprinkled the tribunes with scented water.[13]

A Feast for the Senses

As far back as ancient Egypt, aromatic substances were already being used on festive occasions and in religious rituals.[14] The Romans were also keen on this. On a stroll through the centre of Rome, you would be continuously assailed by clouds of incense from the countless temples. It was made from the resin of different tree varieties, mostly myrrh from the Arabian balsam tree, with its hint of soil and vanilla, and slightly honey-like frankincense from the African Boswellia species.

All perfumes and incense were beyond the pockets of ordinary Romans.[15] But the elites revelled in all the exotic olfactory experiences that money could buy, and perfumed clothes and baths were common status symbols.[16] Banquets were often designed to be feasts for the senses: the tables were frequently rubbed with mint leaves and the floors strewn with flowers to look like meadows. Guests were handed wreaths of roses, violets, hyacinths, laurel leaves and

BLOOD

ODORANT	GROSS FORMULA	MOLECULAR MASS [U]	BOILING POINT [°C]	ODOUR DETECTION THRESHOLD [PPB]
Trans-4,5-epoxy-(E)-2-descenal	$C_{10}H_{16}O_2$	168	272	0.019
1-octen-3-one	$C_8H_{14}O$	126	59	0.2

When blood at body temperature seeps out of a small wound, you won't necessarily pick up any smell at all. But if you pay close attention, you may detect a slightly acidic and metallic aroma not unlike rusted iron or copper. That is the smell of fresh, untainted blood, and it is dominated by an aldehyde called trans-4,5-epoxy-(E)-2-decenal. It has a high boiling point, but since the odour detection threshold is also extremely low, even microscopic amounts will soon be picked up by hunting predators – whether insects, wolverines or sharks – and perhaps also by the werewolves among us. But only when blood is flowing freely, in amphitheatres and field hospitals, say, will the amounts be large enough to be picked up by most people. And then the uneasiness spreads.

Soon, the olfactory experience will be amplified by a ketone called 1-octen-3-one, which is produced when the iron content of the blood reacts with the sebum on the surface of the skin. In addition to a sharper smell of rust, it adds sweetish and fungal components, as well as something reminiscent of the smell of cold meat. Since the boiling point of this substance is much lower than that of the original aldehydes, concentrations will swiftly exceed most people's odour detection threshold.

As the blood becomes darker and coagulates, its smell becomes sweeter and sweeter until it more or less vanishes. If the blood is attacked by bacteria before it has a chance to coagulate – as is the case with menstrual blood in the vagina – it will soon smell rotten, with hints of the cadaverine and putrescine familiar to us as the smell of spoiled meat.

parsley, and in the houses of the most powerful, like Emperor Nero, the air was perfumed with scented water sprinkled from pipes in the ceiling of the banquet hall. It could all reach extraordinary heights of excess: one of the later emperors – the ultra-decadent Heliogabalus – is said to have showered his guests so generously with flowers that several of them suffocated.[17]

In the modern narrative of ancient urban societies, smell is absent.[18] Rome is generally depicted in idealized pastels – all sparkling-clean marble streets and sunshine. At best, we might imagine the warm, omnipresent Mediterranean scents of myrtle, rosemary, jasmine, laurel leaves and resin. Based on habitual expectations, we may perhaps add a dash of freshly washed sheets, olive oil and sundried tomatoes.

But these were irrelevant ingredients in the overall smell that filled the Roman cities 2,000 years ago. And the only people in a position to correct this notion would have been the peasants who constantly travelled into the city to sell vegetables, nuts, wine and meat at the food markets there, or to collect manure from the cess-pits that they could use to fertilize their fields. But peasants don't write history.

The Esquiline Hill

The archaeologist Rodolfo Lanciani was never in any doubt about the stench of corpses, and documented his findings with copious measurements, drawings and photographs. However, later researchers have suggested that his claims about the open mass grave on the Esquiline Hill were, at best, a misinterpretation, and most likely mere bluff and sensationalism.[19]

There is little doubt that Lanciani gave in to the temptation to exaggerate. If there was one thing that struck irrepressible fear into the hearts of nineteenth-century humans, it was the stink of rotting cadavers, which were long considered the cause of all the worst diseases. Lanciani's richly illustrated book about the *puticuli* ran to twelve editions.

Lanciani's later critics also point out the archaeologist's fraudulent ways. Shortly after conducting his excavations, he was accused of illegally exporting artefacts to American museums, and of manipulating archaeological finds for his own personal gain.[20]

The Blind Spots of History

Gaius Maecenas, advisor to Emperor Augustus, ensured that the *puticuli* on the Esquiline Hill were already thoroughly cleaned up towards the end of the first century BCE. Whether they had been open or closed, they were now buried beneath more than 7 metres (24 ft) of soil, after which the land was divided into plots for the villas of the elites.

Lanciani made his discovery 1,900 years later when a sinkhole abruptly opened up between via Carlo Alberto and via Mazzini – now via Cattaneo – while he was conducting an excavation. During his ensuing investigations, Lanciani claimed to have unearthed more than 24,000 cadavers, but no further excavations have been carried out in the area since. That makes it difficult to draw any definitive conclusions. We simply don't know whether Rome was tormented by the stench of corpses from open *puticuli*. Either side may be right.

As in this case, the stories of the everyday smells of the past will always be difficult to spot, because so little has been written down about them. While aspects linked to vision, hearing and taste are extensively documented, the cultural history of smell can only be glimpsed between the lines. All we are left with are the extraordinary scents of perfume and incense. And here, the source material is overwhelming.

3

Heavenly Pastures: The Smell of Salvation

When you step over the threshold of a house of worship, whether temple, mosque or church, you often find yourself in an unusually lofty space – wherever on the planet you happen to be. The creaking of the door as it shuts behind you, the old chap clearing his throat and the occasional rustle of a prayer book all cast barely audible echoes between the walls, very slightly sharper in a stone building. And you become aware at once of a universe of marvellous smells – from wreaths of flowers to censers and perfumed figures – unless, that is, you have the misfortune to find yourself in an olfactorily austere Protestant church.

The Interior of a Church

In the 1860s Charles Dickens conducted systematic expeditions to the Protestant churches in the City of London. He had realized, to his shame, that he was more familiar with the countless temples and churches of Rome than with the houses of worship in his own hometown.

The smells that met him in the darkened naves instantly revived a series of terrible childhood memories: being dragged to church by the hair 'to be steamed like a potato in the unventilated breath of the powerful Boanerges Boiler and his congregation'.[1] Gradually, the impressions became more nuanced. 'There was a remarkably agreeable smell of pomatum,' he says of one church. 'But in other cases, rot and mildew and dead citizens [buried under the floor of the church] formed the uppermost scent.' And in the churches around Mark Lane, 'there was a dry whiff of wheat; and I accidentally struck an airy sample of barley out of an aged hassock in one of them.' As Dan McKenzie once noted about this passage, 'The reader's throat begins at once to feel dry.'[2]

During the Reformation, the censers were removed and the reliquaries packed away. What's more, many of the statues of saints were defaced, their noses often broken.[3] We know little about the motivation for this symbolic destruction of the sense of smell, but the overarching aim was clear: nothing should distract one's attention from God. What we do know is that the Reformation brought an end to an era of spiritual sumptuousness that had prevailed throughout most of the history of religion.

Incense

In ancient times, the Greek gods who lived on Mount Olympus bathed in scents and were even somewhat aromatic themselves – especially the goddess of love, Aphrodite. The early Christians were also passionately interested in smells. A manuscript from the fifth century, *Transitus Mariae*, claimed that Jesus was conceived when the holy spirit penetrated the Virgin Mary with sweet scents.[4] Paradise was pictured as a vast expanse of fragrant flower meadows.[5] And the saved were said to have journeyed to heaven across a bridge of sweet fragrance.

In religion, divinity and humanity were bound together in an olfactory cycle to which mortals had to contribute in order to maintain the connection. Incense, which emits generous clouds of spicy sweetness when strewn on red-hot charcoal, was used in sacrifices to the sun god in Nineveh almost 2,000 years before the Common Era. The ritual was upheld in the temple of Baal in Babylon 1,000 years later, where almost 30 tonnes of incense were burned each year.[6] And we can assume that it was prepared in more or less the same way as in the instructions God gave to Moses:

> And the LORD said unto Moses, Take unto thee sweet spices, stacte, and onycha, and galbanum; these sweet spices with pure frankincense: of each shall there be like weight.
>
> And thou shalt make it a perfume, a confection after the art of the apothecary, tempered together, pure and holy. (Exodus, 30:34–5)[7]

While incense was used in the everyday rituals, animal sacrifices were reserved for more important occasions and could involve anything from birds to bullocks. According to the Assyrian satirist Lucian of Samosata, the Greek gods are ready and waiting, as they 'look down to earth and scan carefully in all directions, stooping down to see if they can see a fire that has been lit, or sacrificial aroma drifting up to them, whirling around with smoke'.[8]

Animal Sacrifices

The procedures for animal sacrifices were strict; here again, Moses received clear guidelines:

> Thou shalt also take one ram; and Aaron and his sons shall put their hands upon the head of the ram. And thou shalt slay

> the ram, and thou shalt take his blood, and sprinkle it round about upon the altar. And thou shalt cut the ram in pieces, and wash the inwards of him, and his legs, and put them unto his pieces, and unto his head. And thou shalt burn the whole ram upon the altar: it is a burnt offering unto the LORD: it is a sweet savour, an offering made by fire unto the LORD. (Exodus, 29:15–18)

Even so, the ritual was often halted as soon the meat was nicely roasted. It still smelt delightful – neither burnt nor charred – and after the celebrants had helped themselves to the best cuts, the onlookers were also allowed to sate their hunger.

Ibn Ezra

Unlike Jews, early Christians weren't interested in animal sacrifices, although they were very keen on incense. Even this enthusiasm soon ebbed, however. After refusing to burn incense before images of Roman emperors, many of them were persecuted and even, in some cases, thrown to the wild beasts at the Colosseum. In Christian circles, those who gave in to the Romans were derided as incense burners.[9]

But there may have been more to it than that. The prominent theologian Clement of Alexandria insisted that the spirit should be elevated above the sensual life of the body, and warned that 'attention to sweet scents is a bait which draws us into the sensual.'[10] The greatest zealots stopped washing and made a virtue of wandering round covered in sweat and filth – just as God had made them.

That stern dogma failed to survive the fall of the Roman Empire and, over the course of the sixth century, incense burning once again

ROAST MEAT

ODORANT	GROSS FORMULA	MOLECULAR MASS [U]	BOILING POINT [°C]	ODOUR DETECTION THRESHOLD [PPB]
Strawberry furanone	$C_6H_8O_3$	128	216	30
Methylpyrazine	$C_5H_6N_2$	94	135	60
2,3-dimethylpyrazine	$C_6H_8N_2$	108	155	20
Acetic acid	$C_2H_4O_2$	60	118	75–1,000

When you place a joint of beef in a hot pan, you set off a chemical reaction between the simpler sugars and amino acids that are the building blocks of the proteins in the animal's body. Known as the Maillard reaction, this not only causes the meat to brown but releases a rich spectrum of odorants: first strawberry furanone, which contributes the sweet, caramel-like and slightly burnt notes, swiftly followed by a couple of pyrazines, which add a slightly cloying nutty aroma, and acetic acid, which brings it all together with a hint of sourness. Some might claim that there is also a waft of flower meadows, damp grain and perhaps a dash of excrement.

The Maillard reaction occurs when you bake bread, roast nuts or coffee beans, and malt barley for beer too; some of the aroma will be retained in the finished products. The odour thresholds are relatively low, and even small concentrations are enough to make most people's mouths water. Unsurprisingly, experiments on rats have shown that these smells also bring down the heart rate and have a calming effect.[11]

Naturally, this is an incitement to manipulation for some: estate agents don't encourage us to bake buns before a viewing just for the fun of it. The aromas created by the Maillard reaction have considerable commercial interest, and synthetic versions of strawberry furanone are lavishly applied to food additives and perfumes.

became a part of religious practice. The spiritual value of the sense of smell was confirmed by both Christian and Jewish theologians, who claimed that the Messiah judged people more by his sense of smell than by sight or hearing.[12] And in the twelfth century, the influential rabbi Ibn Ezra trumped them all. He had no doubt whatsoever about the significance of smell in both the spiritual and the earthly realms: 'The ear is sometimes deceived in hearing sounds, which are only imaginary; the eye, too, sees things in motion, which in reality are at rest; the sense of smell alone is not deceived.'[13]

The Odour of Sanctity

The smells associated with places of worship were supposed to offer a foretaste of paradise – a tiny patch of heaven come down to Earth. It is hardly surprising, then, that those who were closest to God might also smell sweet. People assumed that they were filled with the holy spirit, and that the fragrance itself was proof of their holiness.

When the revered ascetic Father Zossima dies in Dostoevsky's *The Karamazov Brothers*, the monks at the monastery in the fictional village of Skotoprigonyevsk in rural Russia have high hopes of sweet scents. 'When before dawn they laid Father Zossima's body in the coffin and brought it into the front room, the question of opening the windows was raised among those who were around the coffin.'[14] The mere thought of corruption was considered almost absurd, because Father Zossima was undoubtedly a holy man. And the monks' conviction rested on a wealth of traditions: when St Birinus passed away in 649 CE, a scent of lilies had ensued; and it was said of St Hubert – protector of rabies sufferers – that a full century after his death, his final breath had 'spread throughout Brittany an odour so sweet, that it seemed as if God had brought together all the flowers of Spring to symbolize the heavenly sweetness which Hubert would enjoy in Paradise'.[15]

The marvellous fragrance of St Teresa of Avila impregnated the garments and objects she had touched, like tableware and even the water in which it had been washed.[16] These aromas could also persist long after death, as with St Olav – who fell at the battle of Stiklestad in 1030: according to legend, his corpse emitted sweet scents for months. And only after the bones of St Sunniva – who was murdered in a cave on the island of Selja just south of Stadt – were found to give off the sweetest fragrance did it become clear that she too must have been a saint.

Acetone

But the anticipation of the monks in Dostoevsky's novel soon gives way to confusion. 'The fact is that a smell of decomposition began to come from the coffin, growing gradually more marked, and by three o'clock it was quite unmistakable.'[17] Father Zossima's body had undeniably succumbed to the odour of corruption, and speculation began to fly. Perhaps he had not fasted strictly enough: 'He allowed himself sweet things, ate cherry jam with his tea, ladies used to send it to him. Is it for a monk of strict rule to drink tea?'[18]

When the terrible stench continues to grow, pious Father Ferapont is summoned in all haste from his humble wooden cell by the apiary. In an urgent roar, he calls upon the heavenly powers: 'Satan, go hence! Satan, go hence!'[19] But to no avail. And perhaps that's hardly surprising. Later theories about the origin of the odour of sanctity suggest it is caused by hunger and fasting. The resulting increase in the metabolic rate depletes the body's fat reserves, producing acetone, which has a sharp, sweet, fruity and almost flowery scent – precisely like the fragrance that emanates from saints and other holy people in so many accounts.[20] In our own era, we associate the characteristic scent of acetone with everyday products such as nail varnish remover, plastic glue and epoxy paint.

Father Zossima was clearly no puritan, then; but neither was he so gluttonous that he suffered from diabetes, a disease in which long-term insulin deficiency can produce the same sweet scent of acetone. High alcohol use can create acetone too. The question is which of these causes applied in the case of Padre Pio, the Capuchin monk who was canonized by Pope John Paul II as recently as 2002. For one of the conclusive proofs of his sanctity was the scent of violet that always trailed in his wake.[21]

4

A Man's World: The Smell of Condemnation

Miguel de Cervantes' early seventeenth-century novel *Don Quixote* tells the tale of an insane knight errant who travels around rural Spain having adventures with his home-loving squire Sancho Panza. The knight, whose 'sense of smell was as acute as his hearing', remarks early one morning:

> 'It appears to me, Sancho, that you are quite frightened.'
>
> 'Yes, I am,' replied Sancho, 'but what has brought this to your grace's attention at this particular moment?'
>
> 'Because at this particular moment there is an aroma about you—and not of ambergris either.'
>
> 'That may well be the case,' said Sancho, 'but I'm not the one to blame; it's your grace's fault for dragging me out at this inconvenient hour and putting me through these unaccustomed paces.'[1]

The Smell of People

People smell. Our skin steams with sweat and sebum, while our bodily orifices emit a constant flow of breath and intestinal gases. And breath, the exhaust fumes of life itself, is what we first encounter – sometimes even before we've set eyes on the person there in front of us. As long as their oral hygiene is reasonably good, this aroma, which comes from the blood via the lungs, can be sweet, refined and seductive.

The next thing we notice is the smell of skin, first and foremost the substances in the sebum that keep the skin supple – somewhat more acidic behind the ears, faintly sharp and metallic when exposed to sunlight, but still muted and attractive.

Then we become aware of the sweat: most intense in the armpits, between the toes and around the genitalia, where the humidity levels are akin to those of an average jungle. We experience an almost dizzying spectrum of smells, from vinegar, honey, grapefruit, fungus, metal and onion to overripe cheese, rotten meat and cat's piss. Often topped by a remarkably sour, rancid and earthy component that is frequently associated with the stink of a goat. It can get overwhelming pretty quickly.

But if we take a non-judgemental stance, we can venture further into this landscape. The different regions of the body – armpits, feet, neck, scalp, genital regions – all have their own particular aromas. We can sniff our way to them even with a blindfold on, as if they were the rooms in a house.

And just like our homes, each and every one of us has an absolutely individual scent. We smell as different as we look and, in experiments, nosewitnesses have proven to be at least as reliable as eyewitnesses at identifying sexual offenders.[2]

Yet our own fundamental odour is concealed from us, even as it continually shifts as a result of ageing and illness, where we live and what we eat. That said, we do still sense the most abrupt transitions – in our own sweat after exertion or a stressful day at work, for example, not to mention in our intestinal gases, which are a constant source of new surprises.

Olfactory Information

Intestinal gas – or flatulence – is probably what caught Don Quixote's attention. During the expedition, the two companions had constantly found themselves in confrontations with wild windmills, angry guests at an inn and incomprehensible women. Sancho Panza's increasingly frazzled nerves had upset his stomach. Or perhaps it was the smell of sweat. Because stress and fear quickly manifest themselves in a thoroughly distinctive odour that is a match for even the stinkiest fart.

In its simplest form, our sense of smell enables us to locate our fellow humans, the way a newborn baby seeks out its mother's breast.[3] But that's just the start of it. Not only can it identify both diet and age through small nuances in people's breath and body odour, but if we get even closer, we can gain enough information about our neighbour's stress levels and other mood swings to make the polite question 'How are you doing?' largely superfluous.

Unfortunately, all this information acquisition takes place at an unconscious level, and we know little about how it affects our actions. But we will still find that while some people smell repulsive, others may smell neutral or downright delicious. At its most extreme, this can even inspire a profound ecstasy, as it did for Jean-Baptiste Grenouille, the protagonist of Patrick Süskind's novel *Perfume*, which is set in eighteenth-century France: 'her sweat smelled as fresh as the sea breeze, the tallow of her hair as sweet as nut oil, her genitals were as fragrant as the bouquet of lilies, her skin as apricot blossom.'[4]

Stench of a Woman

The Roman satirist Martial made no bones about his opinion of the prostitute Thais:

> Thais smells worse than a grasping fuller's long-used crack [a pot filled with urine], and that, too, just smashed in the middle of the street; than a he-goat fresh from his amours; than the breath of a lion; than a hide dragged from a dog beyond Tiber; than a chicken when it rots in an abortive egg; than a two-eared jar poisoned by putrid fish-sauce.[5]

The Roman Empire was still at its height, and the sauce Martial spoke of was almost certainly *garum*. When he mentions the smell of a crack, he isn't talking about a chamber-pot neatly placed beneath a bed, but one that had been smashed in the street; and the egg isn't just rotten, but contains a chicken. He could hardly lay it on thicker.

Despite the exaggeration, this isn't an entirely unexpected description of women in Western culture, for it forms part of a tradition of mockery that has persisted from the Roman age up to the present day. Older women have tended to be the hardest hit. In the mid-sixteenth century, the young French poet Pierre de Ronsard claimed that their breath alone was enough to set dogs howling and reverse the flow of rivers.[6]

Diabetic and fasting saints were obviously off the hook; but the scent of women *did* occasionally come in for some praise too – and again Ronsard was on the case. One time, he tells us, he was so overwhelmed after venturing into the olfactory sphere of a young girl that he actually fainted because 'your scent fills all my senses with fragrance.' And he describes the most marvellous scent, which

SWEAT

ODORANT	GROSS FORMULA	MOLECULAR MASS [U]	BOILING POINT [°C]	ODOUR DETECTION THRESHOLD [PPB]
Trans-3-methyl-2-hexenoic acid	$CA_7H_{12}O_2$	128	225	1.6
3-hydroxy-3-methyl-hexanoic acid	$C_7H_{14}O_3$	146	265	2–20
3-methyl-3-sulfanyl-hexan-1-ol	$C_7H_{16}OS$	148	220	1.1–1.3
Methanethiol	CH_4S	48	6	0.0002–40
Propanoic acid	$C_3H_6O_2$	74	141	20–170
Isovaleric acid	$C_5H_{10}O_2$	102	176	50–400

We have two types of sweat glands: eccrine and apocrine. The eccrine sweat glands cover our entire body and their sole purpose is to remove surface heat by secreting water. The apocrine sweat glands are located in our armpits and the areas around our genitalia, rectum, navel and nipples. Inactive until puberty, they produce a richer, milky secretion, which increases in tandem with levels of cortisol, a stress hormone, in our blood.

Totally fresh sweat is almost odourless – at best, you might detect a slight tingling sensation of sweet grass. The smell only arises when the sweat is attacked by the bacteria that live on the surface of the skin. Since men and women have different bacterial cultures, their scents also differ: the smell of women's sweat is a tad fresher.

'surpasses in sweetness thyme, jasmine, carnation, raspberry and strawberry'.[7]

But in such cases, the women would ideally be in their teens and certainly no more than their early twenties. They were in trouble if they happened to be menstruating, because that made them nothing short of an environmental hazard. Menstruating women could drive dogs mad, rust iron, blunt knife blades and dull mirrors.[8] The

At first sniff, the smell of apocrine stress sweat is like a cross between meatballs and musty locker rooms, which quickly becomes sharper. More specifically, these are organic acids with a sharp hint of goat mixed with cumin and a sulphur thiol that stinks of onion and cat's piss. The molecules, which contain the hexenoic and hexanoic compounds in the table, have high boiling points. But since the odour detection threshold is low it doesn't take much for the substances to drown out any other smells that may be present.

But when people's nerves are under control, the eccrine variant is the one that will be largely responsible for the smell of sweat, generated mainly by the humid conditions between the toes, in the armpits and the groin, where the concentration of sweat glands is highest. The smell is dominated by broken-down amino acids in the form of the sulphur compound methanethiol, whose boiling point and odour detection threshold are both low. That means only a few molecules are needed before we pick up the sharp stench of rotten cabbage – some would say rotten eggs – with its unmistakable edge of flatulence. As the sweat builds up – without interference from soap and water – the smell will also blend in with sour, rancid propanoic acid and sharply cloying isovaleric acid, as well as a hint of fermentation and cheese. In this case, the odour detection thresholds are a lot higher, so you won't notice the smell until shirts and socks have been soaked with sweat for a prolonged period. The Norwegian term for stinky feet, *tåfis* – or toe fart – is brilliantly apposite.

fourteenth-century English surgeon John Arderne claimed to have observed that the smell of menstruating women could also irritate open wounds, an effect that could be transmitted to the men who had recently slept with them.[9]

A Man's World

Men's sweat smells stronger than women's, especially owing to its piercing acidity, reminiscent of goat.[10] The aversion to the smell of women was probably very much related to the man's position in Western culture. Ever since antiquity, men have had the God-given duty of managing the world. They were best suited to ruling, studying and setting off on adventures, while the women were left to take care of the cooking and childcare, or other roles such as concubine.

The narrative about the female stench was a marker of authority and distance – in much the same way that smell has since been used in the class-based society of the Industrial Revolution, and the multicultural society of our own era, as we will see later. It was deemed an almost innate curse and Martial claimed that nothing could help Thais:

> In order craftily to substitute for such a reek another odour, whenever she strips and enters the bath, she is green with depilatory, or is hidden behind a plaster of chalk and vinegar, or is covered with three or four layers of sticky bean-flour. When she imagines that by a thousand dodges, she is quite safe, Thais, do what she will, smells of Thais.[11]

Witches

In the late Middle Ages, the sexes went head-to-head in a battle between good and evil: the men were soldiers in heaven's army, while the women were servants of the Devil. If a woman failed to acquire a male guardian, she could end up becoming a witch.[12] And there was no coming back from that. A witch could cast a spell on anyone, and if a man was really unlucky, she might even steal his noble parts.

We have good reason to assume that many women's stress and anxiety levels were high during this era – and that will have caused them to secrete the rancid stress sweat. Like a self-fulfilling diagnosis, strong-smelling women were said to be even more tightly bound to evil. And if they did prove to be witches, that repulsive stench would only be exacerbated by their activities – copulating with Satan himself, constantly plundering graves, as well as taking part in executions and gathering around the body parts.[13]

Given these assumptions, witchfinders relied heavily on their sense of smell during the witch trials that took place between the fifteenth and seventeenth centuries.[14] They rarely harboured any doubt:

> [Witches] exhale a stench from the mouth, the whole body, which is communicated to their garments and fills their houses and the vicinity and infects those who approach. We can attribute this to the secretion of a malodorous animal oil, within the organism, arising from the impure ardors which consume them.[15]

One of the favourite poets of the French king, Charles-Timoléon de Beauxoncles, fantasized about a witch who smelt worse than rotting carrion:

> But what was unbearable,
> And what reeked so badly to me.
> Was that the woman and the Devil
> Both stank.

She farts and passes wind for pleasure;
He smelled of old cheese;
And the Devil and the She-Devil
Vied to out-stink each other[16]

These days, given our ocularcentric approach and the Disney cartoons imprinted on our mind's eye, the thought of witches triggers images of crooked noses, protruding eyes, hairy warts and bony fingers. But that's a misunderstanding at best. The truth was that even the most beautiful woman could be unmasked – although the stench that betrayed her may simply have been caused by unease and anxiety.[17]

5

The Devil's Breath: The Smell of Corruption

The Irish author James Joyce grew up in a strict Catholic home at the end of the nineteenth century. In his autobiographical novel *Portrait of the Artist as a Young Man*, he reflects on what awaits him in the underworld:

> All the filth of the world, all the offal and scum of the world, we are told, shall run there as to a vast reeking sewer when the terrible conflagration of the last day has purged the world. The brimstone, too, which burns there in such prodigious quantity fills all hell with its intolerable stench.[1]

He learns about St Bonaventure, who claimed in the thirteenth century that the carcasses of the damned in the depths of this eternal darkness emit such a repulsive stench whenever they writhe in pain that 'one of them alone would suffice to infect the whole world.'[2]

Bonaventure's contemporary, St Hildegard of Bingen, didn't pull her punches either, describing the landscape of hell as 'a long wide marsh filled with the filth and vermin of many types and emitting the worst stink'.[3] And the stench was so penetrating that those who were seduced by the Devil would be infected by it no matter where they were.

Sin

There was no doubt about it: sinful people, men and women alike, stank; according to the fifteenth-century theologian John Wycliffe, lechers and heretics smelt of boars or pigs and 'carried their stink wherever they went'.[4] St Christina the Astonishing of Belgium was so sensitive to the smell of sin that she had to flee to the wilderness or seek refuge from the stench in ovens, trees and towers.[5] Her contemporary, the Franciscan preacher Bernardino of Siena, claimed that the cities were the worst offenders: 'When I enter a city . . . all the good and evil that is done in it comes into my head . . . everything that smells sweet or that stinks.'[6]

Originally set up as humble trading posts along busy rivers, by safe natural harbours and in fertile agricultural areas, towns and cities developed into complexes of intricate districts dotted with churches, castles and military barracks. As in ancient Rome, the inhabitants lived alongside chickens, cows, horses, pigs and other domestic animals, which – along with the rats – defecated everywhere. Besides the obligatory fishmongers and butchers, the workshops and industries remained largely the same: the dyehouses, which fixed their dyes with urine, and tanners, glove- and bag-makers, who used animal dung to make the leather more supple. And all of them had stinking waste to dispose of, especially the butchers whose piles of hair, entrails, dung and blood would otherwise have lain rotting in their backyards.

The streets sloped down towards a central gutter, where all fluids, including the contents of the city-dwellers' chamber pots, eventually combined to form a single stream that was flushed out into the harbours, canals and rivers by sporadic rain showers.[7] When Jonathan Swift, author of *Gulliver's Travels*, wandered through London on a day of drenching rain in around 1700, the smells coalesced into a kind of olfactory map of the city: 'Now from all parts the swelling kennels [gutters] flow/ And bear their trophies with them as they go:/ Filth of all hue and odours seem to tell/ What street they sailed from, by their sight and smell.'[8]

There is little doubt that European cities stank. We are told this by travellers who could identify the odour from many miles away – often well before they could hear the endless tolling of church bells that was typical of the era.[9] But the overall picture was probably more complex.

Perceptions of Progress

Although the people of the Middle Ages and the Reformation may appear to have been far more tolerant of smell than we are today, they were still preoccupied with beauty, comfort and health.[10] Waste barrels were in use, and backyards often had outdoor privies that were emptied on a regular basis. The contents were buried in locations where they would not cause a nuisance. Much of this was dealt with by the city authorities – as in Bergen, where Magnus Lagabøte's City Law of 1276 was amended to incorporate a provision against disposing of waste in Vågen Harbour.[11] Regulations of this sort became increasingly common over the course of the fifteenth century, and violations could be punished.[12]

Although some people still shirked their responsibilities, the mediaeval cities were seldom as full of stink and corruption as many modern interpretations would have us believe.[13] Historians are often herd animals, driven by more or less conscious agendas. According to the German sociologist Norbert Elias, one of these agendas may be to promote a positivist perception of the world to sustain the belief in progress.[14] If we think of ourselves as civilized, the implication

– at least if we leapfrog nimbly over the idealized city societies of antiquity – must be that we used to be uncivilized. This narrative naturally works best if the gloomy and foul-smelling mediaeval cities with their bottomless mud and endless thunderstorms slowly but surely evolve into gleaming white, urbane and modern metropolises where the sun always shines.

We can see something of the same tendency in modern novels, where stench is consistently used as literary shorthand for the primitive and the backward – like in Booker prize-winner Marieke Lucas Rijneveld's *Bildungsroman* set in rural Netherlands, *The Discomfort of Evening*, with its grotesquely detailed enumeration of the stink of stewed udders and animal dung.[15]

The Devil's Breath

The people of the late Middle Ages did, of course have far richer olfactory experiences than we do today, but there is little to suggest that they saw it as a problem.[16] Until, that is, the arrival of the plague pandemics, when the cities were hardest hit. Much of the harmony gave way to panic.

After the Black Death abated, the most intense waves of plague came in the sixteenth and seventeenth centuries. Few doubted that they were caused by God's wrath over a sinful world. Since it was the Devil's job to punish humanity, the populace must do its utmost to avoid any contact with what became described as the devil's breath. The medical establishment was more or less united behind this concept and not even a reformer as thoughtful as Martin Luther called it into question: it was the Devil who 'poisoned the air or otherwise infected the poor people by their breath and injected the mortal poison into their bodies'.[17]

BREATH

ODORANT	GROSS FORMULA	MOLECULAR MASS [U]	BOILING POINT [°C]	ODOUR DETECTION THRESHOLD [PPB]
Acetone	C_3H_6O	58	56	20,000–65,000
Methanethiol	CH_4S	48	6	0.0002–40

An exhalation contains 0.5–1 litres of air, drawn from the depths of your lungs. In addition to carbon dioxide, it carries the volatile by-products of bodily processes. It can have a strong smell, but since respiration is so intimately linked to your organ of smell, you won't experience it yourself. You have to rely on information from your nearest and dearest.

After a night without food, your breath will contain fruity notes of acetone caused by the metabolism of your body fat. That will be supplemented by molecules of the food and drink that have entered your blood. Onion will contribute sharp sulphur compounds, alcohol a hint of fermented fruit, and red meat a whiff of fish that has been lying around for too long. The more you've consumed, the stronger the smell.

That said, the most important factor influencing the olfactory impression is usually the state of the oral cavity itself. Residues of food and drink accumulate mostly between the teeth and on the rough part at the very back of the tongue. The surfaces are densely packed with bacteria, and the high temperature and humidity provide ideal conditions for their further decomposition into odorants. Specifically, we're talking about acids that smell of sour milk, amino acids that stink of spoiled meat and sulphur compounds with a stench of rotten cabbage. Sulphur compounds tend to dominate, especially methanethiol, which is also present in the smell of sweat. The odour detection threshold is so low that even small doses are enough to cause discomfort.

Controlling Contagion

By as early as the second century CE, the Roman physician Galen had already claimed that the plague was caused by an imbalance in the composition of the air originating largely from the products of decomposition and decay.[18] During the Renaissance, that theory was adapted to reflect the Christian conception of the world, and now the dangerous stench came not from the heavens but from the underworld, from hell. And conditions were worst in the vicinity of cemeteries, bogs, swamps and stagnant water.

Not only should people steer clear of such places, but they should shun the inhabitants. If this was unavoidable, they should wear garments that left no skin uncovered. And the body, which was seen as both porous and absorbent, must be sealed with oily salves – including the anus.[19] Even with all these precautions, people should still stick to the middle of the street to avoid coming into contact with people leaving their houses, and above all to avoid their foul-smelling breath.[20] In *A Journal of the Plague Year*, Daniel Defoe – better known for his novel *Robinson Crusoe* – writes of the dilemma in which his narrator's physician friend, Dr Heath, found himself. 'Who durst smell to that breath for his information, since, to know it, he must draw the stench of the plague up into his own brain, in order to distinguish the smell?'[21]

Antidote

Since the smell was worst in the cities, which were also hotbeds of sin and contagion, the countryside soon came to be seen as healthy and unsullied. During the various plague pandemics many European aristocrats fled to remote hunting lodges, and estates were built overnight – behaviour familiar to us from other pandemics, not least our very own COVID-19.[22]

Those left behind had to put up a fight. Some barricaded themselves into their homes after stocking their cellars with food and drink, determined to sit tight until the plague had passed.[23] Others tried to combat the contagion with even worse smells. Physicians recommended starting the morning with a few deep breaths of the air in the privy, preferably on an empty stomach. Later in the day, they advised drinking urine and sniffing rotten cheese. If you could afford a smelly goat, your home could be considered reasonably safe.[24]

Similar strategies were implemented in public spaces, and were taken to the extreme in Poland, where heaps of rotten animal carcasses are said to have been piled up on the streets and in town squares.[25] Elsewhere, large bonfires burned day and night, often fed with feathers, sulphur and human hair.[26] Even burning old shoes was better than nothing.[27]

Sweet Smells

After a while, people also started using pleasant smells to combat the devil's breath, which – unsurprisingly – contained the rotten effluvia of the pus, decomposed tissue, vomit and diarrhoea of the sick. Many doctors equipped themselves with beak-like breathing masks, which they filled with aromatic herbs, not only for the obvious purpose of alleviating the stench but in the hope that it would have a certain cleansing effect.

The sober-minded narrator of Defoe's *A Journal of the Plague Year* always carried around 'preparations of strong scent', just to be on the safe side.[28] He tells of churches that were like 'smelling bottles' owing to the blend of different aromas.[29] People were told to strew rose leaves and lavender on the floors of their homes, and the French physician Jean de Lampérière advised lighting aromatic bonfires at sunset, using wormwood, tansy, rosemary and sage. Even the firewood should be of a type that could withstand rot and was therefore further from hell – such as juniper, ash, walnut, willow and pine.[30]

The seventeenth-century English courtier and natural philosopher Sir Kenelm Digby soon developed a handy aromatic tobacco, which contained not only exotic spices, such as nutmeg and cinnamon, but rarer ingredients, too, like ambergris – a secretion from the stomach of the sperm whale – and musk, which was extracted from the sex glands of the male musk deer.[31] The battle against the plague had stimulated the quest for new scents, but these were rare and costly items, reserved for wealthy aristocrats and the clergy.

6

The Musk Deer: The Smell of Deception

It has long been unclear whether the Italian explorer Marco Polo actually made it to China at the end of the thirteenth century or whether the entire business was a fantastic fib. In his journal, we read of his experiences along the Silk Road, with much talk of precious stones, pearls, spices and silk. But he also writes about that legendary substance, musk; and the wealth of detail he provides on this precise subject may well be among the best proof that he actually *did* reach China.[1]

Musk

The scent of the faintly blushing, waxy and slightly grainy substance that is musk is a complex affair: it starts off as an intense animal aroma that many perceive as cloying to the point of nausea. At the same time, there are surprisingly light notes of fig and nuts, which we also associate with the scent of healthy human skin. The substance was prized by the ancient Romans and Greeks. The early Christians claimed it was the fragrance of the Garden of Eden itself, while Muslims imagined that the mountains and earth of paradise, and even the heavenly virgins – the Houris – were made of musk.[2]

Yet little was known about the substance, although people assumed that it was drained from the navel of a mysterious animal that lived in the mountains far out towards the eastern edge of the world. Marco Polo was able to correct this misapprehension. He claimed that it was a secretion from the external sex glands of the adult male musk deer, which was 'shaped like an antelope, hoof-tailed antelope, hairy deer and thicker, without horns, [*sic*] Mouth has four teeth, two on top and bottom, white as ivory'.[3] If the description of the animal was somewhat vague, Marco Polo was able to offer detailed information about the extraction process.

> Hunters hunt this beast when the new moon rises, which is also the time when it excretes musk. After catching this beast, cut its blood bag. Put it in the sun to dry. Its strong fragrance . . . is from the blood in the bag.[4]

And he confirmed that the meat also tasted fantastic.

Cats and Beavers

Given the scarcity of routes to the east, musk long remained the preserve of kings and nobles, but once trading opened up along the Silk Road during the sixteenth century, the attractive substance became available to more of the elite. But the fact that even tiny amounts continued to sell for exorbitant prices spurred an entire industry of imitations. Sometimes genuine musk was adulterated in a process involving goat bile, hot ashes and the ground-up droppings of black pigeons, and recipes for entirely counterfeit products were also in circulation.[5] Only a small and exclusive minority had ever smelt the original scent, so people were easily fooled.

MUSK

ODORANT	GROSS FORMULA	MOLECULAR MASS [U]	BOILING POINT [°C]	ODOUR DETECTION THRESHOLD [PPB]
Muscone	$C_{16}H_{30}O$	238	328	10–60
Muscopyridine	$C_{16}H_{25}N$	231	338	5–20
Civetone	$C_{17}H_{30}O$	250	342	0.3–10

If you ask for a drop of musk on the back of your hand at a perfumier's, it's the synthetic version you'll be offered. Although there is no comparison with natural musk, which contains more than fifty different compounds, you can still expect to pick up the most essential components. First and foremost, muscone, a sweet ketone with animalic notes. You will also detect a pyridine derivative called muscopyridine, which supplements the animal character with a hint of urine and sweaty feet. In civet oil, muscone is replaced with civetone. Other than that, the substances smell pretty similar, and they are separated by only a single carbon atom. Their high molecular mass ensures that these aromas will always be unusually rich. Many would say that they quickly become almost overpowering.

Since musk compounds have high boiling points and their molecular weight is very close to the threshold beyond which evaporation is impossible, these smells will linger for a long time. At the same time they have low odour detection thresholds. These properties make musk compounds the ideal backbone for almost any perfume, and most products currently on the market are blended with larger or smaller amounts of synthetic musk – not least Chanel N° 5, the best-selling perfume of all time.[6]

Yet many of us will associate the smell of musk mostly with detergents, laundry products, shampoos and hand soaps. This is because large amounts of synthetic musk are often added to such products to mask other smells and offer a lasting impression of cleanliness.

Nonetheless, there were decent replacements for authentic musk, the most important of which was civet oil – a butter-yellow glandular secretion the African civet uses to mark its territory. A far cheaper alternative, castoreum, could be extracted from a gland between the external genitals and anus of the common beaver. When dried and ground into a rough powder, castoreum has hints of leather, smoke and tar, with undertones that vary depending on the beaver's sex: the female's secretions often have a faint hint of blue cheese, while those of the male are more reminiscent of motor oil. The substance was more commonly used as medicine than a perfume and was potent enough for Norwegian grimoires of the seventeenth century to recommend it as an effective means of combating evil spirits.[7]

Synthetic Musk

One critical aspect of the natural musk variants was ease of access to the raw materials. While civets could be kept in captivity and their territorial markings collected each morning, both beavers and musk deer had to be killed to obtain the sought-after secretions. No fewer than 140 animals had to be sacrificed to obtain a single kilogram from the musk deer, a fact that almost led to its extinction.

In light of all this, it was nothing short of sensational when an experiment in the dynamite industry in the late nineteenth century led to the accidental discovery of a substance dubbed nitro musk. True, it was a mere imitation, with a scent more reminiscent of pomade and old hairdressing salons than genuine musk. But a spark was lit and half a century later the Croatian chemist Lavoslav Ružička produced the first entirely synthetic musk, winning the Nobel prize for his pains. New variants followed, with names like nirvanolide and helvetolide.[8] Like their forerunner, all were based on chemicals from the petrochemicals industry.

Porous Bodies

For anyone wishing to reduce their body odour and avoid smelling like an emissary of the Devil, better hygiene was an obvious alternative. Regular washing had been common practice since antiquity. The Romans bathed frequently, the wealthiest sometimes several times a day. But norms changed over the course of the late Middle Ages. Influential physicians had observed that dirty bodies were healthier than clean ones.[9] Bathing, which further opened up the already porous body to the atmosphere, was soon deemed both dangerous and foolhardy.[10] Even clothing shouldn't be washed too often.[11]

In other words, more scrupulous attention to hygiene was not an option when the ravages of the plague were at their height in the sixteenth and seventeenth centuries. And this perception held sway almost up until the present day, largely irrespective of class. Even the royal palace built in Kristiania (modern-day Oslo) in the nineteenth century did not have a bathroom.

A quick scrub with a damp cloth was all people needed, possibly with some added scent. Hair could be cleaned with a light starch- or clay-based powder, also scented. And bad breath could be banished with perfumed candies or by chewing aniseed and aromatic herbs.[12]

Perfumed Generals

Undesirable body odours could also be camouflaged with a veneer of other smells. Like musk, these were often animal in origin, such as beeswax, and were cheap enough even for the pockets of the petite bourgeoisie. The blends could also be complemented with plant products, like the formula proposed by Louis Guyon, advisor to the King of France. He claimed that the terrible stench of women could be eliminated with 'a dab of spike lavender or sweet almond oil mixed with a few grains of civet or musk on a midwife's finger to be placed as far up the vagina as possible'.[13]

The target market for perfume was primarily foul-smelling women, but after the plague pandemics died down in the late seventeenth century, men also became involved. Almost overnight, the use of perfume became associated with virility and masculine strength, and the city streets teemed with powdered Adonises wearing perfumed gloves and tall wigs, wafting clouds of fragrance.[14] And they even went to war like this. Well into the eighteenth century, French generals turned up on the battlefield heavily powdered and perfumed.[15]

Floral Scents

Among the urban bourgeoisie, the use of perfume soon became accepted practice for both men and women. At the same time, though, the popularity of animal fragrances waned.[16] The far lighter, floral, spicy and fruity aromas were better suited to more refined noses. Lemon, orange, lime, bergamot and, above all, jasmine soared in popularity to become the most sought-after ingredients.

Plant-based scents can be found in petals, stems, resin or roots, and are extracted by cutting, squeezing or crushing. To produce the perfume, they are then infused in alcohol, water or wine to make essences that perfumiers blend into cohesive compositions. One of the first such blends was eau de cologne, which was produced by dissolving essence of rosemary and lemon in grape spirit; thousands of variants have since evolved.[17]

Manipulation

Our smell defines us. All of us have our own entirely personal aroma, determined by age, sex, lifestyle and mood swings.

We may use perfumes simply because we like the scent, but we may also want to smell a bit different to the way we actually do. We construct a personality that better accords with the way we wish to be perceived – whether by God, Satan or our fellow mortals. Men generally douse themselves in musk-based products to appear more masculine, while women enhance their femininity with roses, orchids, lilies, citrus fruits and vanilla. The scent of lavender evokes extraordinary cleanliness in an age when this substance is largely used in all kinds of soaps and laundry products, whereas coconut oil stirs thoughts of exotic South Sea adventures on even the coldest winter day.

Unsurprisingly, the puritans of the Reformation sternly advised against the use of perfume, even when the plague pandemics were raging. They claimed that the artificial sweetness concealed humanity's innate corruption. And bodies that were 'so perfumed and bathed in odoriferous waters . . . must one day be throwne (like stinking carion) into a rank & rotten grave'.[18]

And they weren't the first people to think along these lines either. A full five hundred years before our modern era, the Greek poet Solon – also considered one of the fathers of Athenian democracy – introduced laws prohibiting trade in perfume. And the philosopher Socrates claimed that the only scent appropriate for men was the massage oil used in gymnastics.[19]

Many ancient Roman intellectuals also scorned the use of artificial fragrances. According to the playwright Plautus, it could have particularly serious consequences for older women: 'When their perspiration mixes with the perfume, it's just as if a cook has poured all his soups together. You'd never know what it is they smell like – but the stench is awful.'[20]

Commercial Control

With the exception of a few brief periods, perfume has been the sole preserve of women, and this situation has persisted all the way up until the present day. In recent decades, though, we have seen a shift that is reminiscent of the more relaxed attitudes of the seventeenth century. Men are back in the game, almost certainly pulled in by massive ad campaigns by the perfume industry, but perhaps also by a general decline in self-esteem. The wealth of new synthetic scents may also have played a role: the perfumiers' arsenal now contains more than 1,800 different aromas, and new ones are constantly being launched.[21]

When we speak or write about smell these days, we are almost exclusively referring to perfumes. The Canadian cultural historian Constance Classen finds this troubling:

> To limit olfactory aesthetics to the productions of the perfume industry . . . would be like limiting the visual arts to the field of fashion design. There is a whole world of vital olfactory imagery and meaning which cannot be, and is not meant to be, encompassed in a perfume bottle.[22]

7

The Liberated Gut: The Smell of Happiness

In the twelfth century, Roger Prior of Durham was visited by a demon, which entered his bedroom and let fly such a powerful fart – or pump, toot, trump or flatus if you prefer – that the stench lingered for almost three days.[1] You'd hardly expect less from an emissary of Satan who was undoubtedly well versed in the art. In Dante's *Divine Comedy*, published in the early fourteenth century, Barbariccia, the leader of a gang of demons, even issued orders via his rear end.

> Along the left-hand dike they wheeled about;
> But first had each one thrust his tongue between
> His teeth towards their leader for a signal;
> And he had made a trumpet of his rump.[2]

Fart

In addition to the stench of corpses and corruption, the hell that people pictured in the Middle Ages was also liberally supplied with the stink of intestinal gases. This is clearly linked to a taboo in the world of mortals, especially among the elites. When the mighty Viking chieftain Einarr Thambarskelfir dropped a fart during a

FAECES AND FLATULENCE

ODORANT	GROSS FORMULA	MOLECULAR MASS [U]	BOILING POINT [°C]	ODOUR DETECTION THRESHOLD [PPB]
Methanethiol	CH_4S	48	6	0.0002–40
Dimethyl sulphide	C_2H_6S	62	37	1–20
Hydrogen sulphide	H_2S	34	60	1–130
Skatole	C_9H_9N	131	265	0.2–2

Intestinal gases are produced as a result of fermentation in the large intestine and are, in principle, a foretaste of what follows: excrement, consisting largely of undigested fibre, dead bacteria and gut cells that have been shed. However, there is a considerable difference between the two when it comes to both chemistry and smell.

In addition to carbon dioxide, hydrogen and nitrogen, intestinal gases contain a small proportion of highly volatile sulphur compounds – the sources of the smell. Despite some variation in composition, methanethiol, with its odour of rotten cabbage (also found in heavy sweat and halitosis), leads the way, along with dimethyl sulphide,

banquet at the home of his rival Harald Hardrada – albeit while in a drunken stupor – his resulting sense of shame was so devastating that it culminated in murder and disruption of the royal line.[3]

Farting is an unavoidable consequence of our body's metabolic processes, and – like it or not – we all fart at least twenty times a day. The fart contains molecules from any undigested food and worn-out cells that happen to be in our digestive system – all accompanied by a distinct, sour whiff of sulphur.

Towards the middle of the last millennium, the taboo against farting faced ever more frequent challenges. Eustorg de Beaulieu, a sixteenth-century French poet, scorned the self-important souls which contributes a more cloying, boiled cabbage smell, as well as a hint of seafood. This is, of course, an important component in the smell of the sea too. Hydrogen sulphide, with a distinctive aroma that is often associated with rotten eggs, also makes a minor contribution.

However, these intense sulphur compounds quickly evaporate from fresh faeces and are replaced by the nitrogen compound skatole, whose name is derived from the Greek word *skato* meaning excrement. This substance results from the decomposition of amino acids linked to the consumption of protein-rich foods like oats, legumes, milk, fish, meat and eggs. Given its high boiling point and low odour detection threshold, only small amounts are needed – if this is any consolation – before you will notice the smell, which is generally described as faecal. At the same time, it is often perceived as cold, complex and relatively rich, and some people will pick up hints of coal tar and beetroot.

In very tiny doses, however, skatole can have a pleasant smell of flowers – jasmine and orange blossom to be precise, both of which contain minuscule amounts of the substance. That is why skatole is widely used in the perfume industry.

who clenched their buttocks, and praised the bold and forthright fart that positively seethed with a lust for life:

> A silent gas scented like a shepherdess's fart.
> A fart perfuming a laundry-maid's twat
> Windy gas, and you, violent fart
> That many ladies laugh as they let fly beneath their nose
> Stiff, impetuous gas that blows
> Sometimes watery, sometimes muddy.
> Fart bouncing high and clear without discord
> Always joyous, singing until death

You are both immutable in will
And unassailable in power and force.[4]

Beaulieu's contemporary, the humanist and scholar Erasmus of Rotterdam, declared that people enjoyed the smell of their own farts.[5] In his book *De civilitate morum puerilium* (On Civility in Children), a guide to Christian mores for the bourgeoisie, he listed prohibitions against various minor infractions – such as spitting out your food and picking your nose at table. But when it came to farting, he took a softer line: let the wind fly free – because keeping it in could make you ill. However, the noise should be camouflaged with a cough.[6]

Faeces

An experiment conducted in the 1980s to discover what two-year-olds were and were not prepared to put in their mouths found that fewer than one-third could deal with a small grasshopper, while just one in ten were willing to chew on a lock of human hair. By contrast, more than 60 per cent were prepared to polish off a spot of fresh faeces.[7] And there is much to suggest that Western children reach close to school age before the aversion to human faeces is fully established.[8]

Even in the adult world, the boundaries used to be more relaxed than now. The people of Renaissance Europe had a higher tolerance not only for farts but for faeces too – generally more so than for the smells of sweat and bad breath.[9] A similar liberality was extended to urine. One story tells of an inn with gutters on the floor where the urine could flow freely.[10] Admittedly, the pale, clear liquid has only a faint smell of salt to begin with. But when it matures or if you've let too much time pass since you last urinated, it will darken into a brownish-yellow colour and give off a sharp smell of ammonia vapour together with a group of rancid organic compounds.[11] And the older you are, the more intense the smell, especially for men.

Behind the Curtains

France's Sun King, Louis XIV, was known for his bad breath and for having bathed only twice in his entire life – and both times it made him sick.[12] In 1661 he embarked on the construction of the Palace of Versailles, just southwest of Paris. Built of natural stone and brick, it contained 350 apartments, salons and a hall of mirrors, and was set in a landscape garden with meticulously clipped hedges, sculpture parks and hundreds of fountains. The building could house 5,000 guests and every aspect of its construction was executed to the highest possible standards – with one single exception: the toilet facilities.

Like most European castles, no plans were made for handling urine and faeces. Some people used chamber pots while others relieved themselves behind the curtains, in stairwells or fireplaces, which were regularly cleaned by an assiduous army of servants. Although both guests and court were thoroughly doused in perfume, the stench could obviously become intense at times. Yet there is little to suggest that this was seen as a problem; indeed, several ambitious aristocrats ordered their own servants to urinate around their mansions to emulate the same aroma.[13] The massive amounts of urine that seeped down into soil and continued to ferment there must have created a stink that we modern folk will only occasionally encounter on packed trains and in vintage waiting-room urinals.

Unfortunately, not everyone was equally robust, even in the days of Louis XIV. In October 1694, the Sun King's sister-in-law, Liselotte von der Pfalz, complained about defecation to her aunt Sophia

Electress of Hanover, and was told in no uncertain terms to pull herself together.

> It indeed seems that you . . . know nothing of the pleasure of shitting; that is the greatest of your misfortunes . . . for it can be said that of all the necessities nature has subjected us to, that of shitting is the most agreeable. One finds few people who shit who do not think their turds smell good.[14]

Puritanism

In the early sixteenth century, the French humanist François Rabelais was an early adopter of a kind of philosophy of freedom: follow your appetites and do what you will. Only nature sets the limits. This new openness also encompassed sexuality, setting generous boundaries for what was deemed coarse, obscene and indecent.[15] Rabelais thereby introduced an acceptance of the human body that would survive both the Reformation and the plague years.

Towards the end of the seventeenth century, however, all this began to change. The French Academy of Sciences embarked on a systematic clean-up of the wealth of scatological vocabulary in the French language.[16] And in the century that followed, the demand for cleanliness rocketed. To be a spreader of stenches – not least farts – was to exile oneself from polite society. 'Repression related to the anus and the urinary tract was now widespread among the higher echelons of society,' who flocked to their physicians seeking help for their agonizing neuroses.[17] Here we can see the seeds of the golden age of psychoanalysis that would follow a mere century later.

Taboo

Although excrement and intestinal gases are a significant part of our existence, they have become taboo over the past two centuries. Neither Ibsen, Bjørnson nor Hamsun ever mentioned them in writing. And a search in the database of Norway's National Library for fart-related words such as *fis* and *flatulens*, along with relevant synonyms – including the Norse word *freta*, just to cover all bases – reveals that they are largely absent from the entire literary output of the twentieth century. The rare exceptions include some manuals of veterinary medicine and the occasional silly kids' book published towards the very end of the period. We also get a glimpse of the subject in a journal written at Grini Prison Camp during the Second World War. There is little better way to undermine rigorous discipline than a forthright, well-timed fart – and all farting during roll-call was punished.[18]

Under Hitler's regime, extreme hygiene in all its forms was a principle in its own right. So it must have been a major problem that the Führer himself had problems with flatulence, especially when under stress. It was often downright difficult to hold dinner parties, and it didn't improve matters when he switched to a vegetarian diet. Even extensive treatment with Dr Köster's strychnine-based anti-flatulence tablets failed to have any noticeable effect.[19]

Fart Jokes

'If you can make a joke in an area where there's anxiety, a little bit of taboo, then you get huge laughs,' says British comedian John Cleese, who made ample use of fart jokes in the *Monty Python* films.[20] And the same furrow has been successfully ploughed by many other comics, including Louis de Funès, Jim Carrey, Leslie Nielsen, Mike Myers and Eugene Levy.[21]

But the king of flatulent comedy was Le Pétomane, the fart artist who more or less ruled the French entertainment scene in the late 1880s. Joseph Pujol, as he was actually called, had discovered that he had the ability to fart on demand and in limitless quantities. By bending over with mouth closed and nose pinched shut, he could build up massive overpressure in his lower gut, which allowed him to achieve uninterrupted farts that lasted up to fifteen seconds. He could also vary their pitch and volume by controlling his sphincter, and this enabled him to produce simple melodies, including the French national anthem.

Dressed in a 'coat, red breeches white stockings and gloves' he generally opened his show with the short, sharp toot of a little girl, the meaty fart of a mother-in-law and the bashful trump of a bride on her wedding night. He would then go on to imitate owls, pigs, frogs and bees. And even a dog with its tail trapped in a door. People laughed themselves silly and in a matter of years, he was France's highest-earning entertainer.[22]

Of course, Le Pétomane's farts didn't smell. They were based on fresh air, untainted by milk, oats or pork. Yet the gleeful association alone was enough to drive his audiences wild.

8

Enter the Eye-Man: The Smell of Reason

By the sixteenth century many European aristocrats had already acquired comfortable rural retreats where they could escape from the plague-infested cities. Although the plague had gradually loosened its grip, this trend suddenly enjoyed a revival in the eighteenth century. From the England of the Hanoverians to the Russia of the Tsars, countless pleasure palaces and towering dachas sprung up, surrounded by extensive parkland. Like a kind of microcosm of nature, these grounds contained everything from overgrown forests interspersed with glades and lakes to well-tended fields with gambolling livestock, cooing wood pigeons and the occasional plaintive nightingale; but most importantly of all, there had to be an eccentric hermit, a recluse, at the centre of it all.[1]

The Noble Savage

The hermit – who, by definition, had withdrawn from social intercourse – lived in a cave, beneath a root or in a rustic ruin, ideally one that had been designed by a prestigious architect. And in the front garden where moonlight and sunbeams could barely penetrate the foliage, there would generally be a little table complete with reading glasses and a weighty, learned tome.[2] The hermit himself was usually a barefoot old-timer – the older the better – dressed in a shepherd's smock with untrimmed nails and an untamed beard. Here he would spend his days in solitude and meditation, like some unkempt Diogenes, the legendary everyday philosopher of ancient Greece who lived in a barrel beside the marketplace in Corinth. But although the hermit in full regalia appeared to be an authentic sage, he was only there to be observed from a distance – like an ornament or a backdrop. There was no need for more in order to communicate the message that the owner of the estate was in step with the spirit of the age. And it went in only one direction: back to nature.

Red-Leaf Rose

In their spiritual strivings, hermits were meant – like Diogenes – to be above any signs of worldly cleanliness, and their abodes must have had an acrid, rancid smell of humans. We can imagine that, when supplemented with the reek of soil and mildew from the earthfast constructions, this added up to the same luxuriant olfactory universe that had characterized the countryside for thousands of years. We are back on the Icelandic farm.

But that wasn't what the aristocracy were after, much less what they paid for. What they wanted was an immaculate, clean nature, miles away from mud and dung, miles away from damp cows and dirty human bodies, miles away from the septic tanks in both city and countryside. They wanted to indulge in the scents of dewy flower meadows, summer rain, babbling brooks and waterfalls. They wanted to be enraptured by the crisp night air at dusk and dawn; by the prickling chill of crystal-clear mornings; by intense rain showers that stirred up exquisite scents of elderflower, water hemlock and

SOIL

ODORANT	GROSS FORMULA	MOLECULAR MASS [U]	BOILING POINT [°C]	ODOUR DETECTION THRESHOLD [PPB]
Geosmin	$C_{12}H_{22}O$	182	270	0.4

If you are lucky enough to come across a newly ploughed field with gleaming black furrows on a sunny spring day, you will pick up a distinct and acidic smell that many would describe as fresh, rich and earthy with a gentle hint of mould. But this is only a foretaste of what awaits you if you go further, into the forest. Here, even the smallest opening in the soil is an almost inexhaustible source of olfactory experiences: from bright tones of camphor, coconut fibre and anchovies in one place to overripe peaches, dried orange peel and wet fur in another. It is almost impossible to predict what lies in wait, but the material beneath the trees with the shallowest root systems, such as birch and pine, appears to be somewhat more diverse and complex.

Various different molecules come into play but the terpene compound geosmin will be one of the chief ingredients. It smells of raw potato, slightly mossy with a hint of mould. The molecule is medium-sized and is thought to be an important signal substance in mycorrhiza – the symbiotic exchange of nutrients between fungi and plants in the underground community.[3] And once you know the smell, it becomes almost like eavesdropping on the silent communication between them.

Although geosmin has a high boiling point, its low odour detection threshold means that you can pick it up even on cold days. And its release becomes increasingly rapid in humid conditions, which contributes to the extraordinary, intense smell of sudden summer rain after a period of drought. For many of us, this has become the smell of summer itself.

Geosmin will linger in certain vegetables, including the earthy tang of beetroot, and this substance is what accounts for the taste of corked wine. Geosmin can also occur in fresh water and freshwater fish, where it is assumed to indicate poorer quality, although the levels will rarely be high enough to be harmful to our health.

red-leaf roses. And, if you were a woman, this euphoric synthesis of smells could even – according to the experts – end in orgasm.[4]

A small selection of agricultural smells was also admissible: freshly cut grass, newly ploughed fields, dried hay, sun-warmed grain and fresh apples arranged on a blanket in the sun. Madame de Pompadour, mistress of Louis XV, had several country estates where, in addition to aromatic beds of thyme, lavender, lily and marjoram, she grew her own vegetables and played shepherdess with newly washed sheep.[5]

Rotten Apples

The French philosopher Jean-Jacques Rousseau was the spokesperson for the new wave. After a prolonged era of plagues and crushing religiosity, he boldly declared that people were born free. But that freedom could only be fully realized in the state of nature, he claimed, and civilization had now robbed us of that possibility: 'Man was born free, and everywhere he is in chains.'[6] Even so, it was still possible to strengthen one's connection to nature, primarily through greater reliance on one's immediate sensory experiences. And he lauded the sense of smell as one of mankind's 'earliest teachers of philosophy'.[7] It was able to reach the very roots of life and could touch the soul far more deeply than either sight or hearing.[8]

Rousseau argued for the individuality of sensory experiences, in which olfactory preferences in particular were shaped by our personal history. On one occasion when the German writer Johann Wolfgang von Goethe was waiting at the home of his colleague Friedrich Schiller – who, like him, was greatly inspired by Rousseau's ideas – he realized:

> an air that was beneficial to Schiller acted on me like poison [. . .] I had not been seated long before I felt a strange indisposition steal over me, which gradually increased, until at last I nearly fainted. At first I did not know to what cause I should ascribe this wretched, and to me unusual, state, until I discovered that a dreadful odour issued from a drawer near me. When I opened it I found, to my astonishment, that it was full of rotten apples. I immediately went to the window, and inhaled the fresh air, by which I was instantly restored. Meanwhile his wife came in, and told me that the drawer was always filled with rotten apples, because the scent was beneficial to Schiller, and he could not live without it.[9]

Enlightenment

Rousseau was also one of the social philosophers who most inspired the French Revolution. He had little time for the decadent aristocrats who indulged in a narrow adulation of nature in his name, claiming that the paternalism of the monarchy, nobility and clergy must be dismantled entirely if humans were to live spontaneously once more, with a common will to good – as in the state of nature.

Many of the philosophers of his day shared Rousseau's aims, but they increasingly began to distance themselves from his romantic glorification of nature, declaring that future social development must be based on science and reason. One such philosopher was Immanuel Kant, who lived in Königsberg – now Kaliningrad – a city of 50,000 situated in the delta where the Pregolya River flows out into the Baltic Sea. Kant, unlike the impulsive and somewhat hyperactive Rousseau, was most decidedly a man of routine. Once the fob-watch became available, he structured his days down to the tiniest detail, dividing them between work, short walks and meals. And at supper, he drank half a bottle of wine – never more, never less.

He lived alone, travelled nowhere and was utterly ignorant of the salt and stormy smell of the Atlantic Ocean, the refreshing dash of glacier in the Alpine air, the spicy aromatic scents of the Mediterranean and maybe even the rich bodily emanations of an intimate partner. Perhaps the fact that he lived out his life inside his own lukewarm bubble is what inspired him to conclude that smell was the most dispensable of our senses: 'It is not worthwhile to cultivate or refine it for the sake of the pleasure we can get from it; for disgusting odours always outnumber pleasant ones (especially in crowded places).'[10]

As if that wasn't bad enough, the sense of smell also helped muddy the rational basis for modern scientific thinking.[11] And here, Kant was completely in agreement with Aristotle and the rest of the natural philosophers of antiquity. They, too, soon deemed the sense of smell deficient because it was too diffuse and impossible to categorize. They had of course noted that smells increased with rises in temperature and humidity. But they took it no further. And they archived the sense of smell under the function of respiration, considering that it had little else to contribute to knowledge about the world.[12] But they never understood it completely and admitted as much.[13]

On the other side of the Baltic Sea, the Swedish naturalist Carl Linnaeus – whom Kant held in high esteem – was making one last attempt to salvage the remnants. After classifying the entire plant and animal kingdoms in his system – species, genus and all – he tried his hand at smell. He ended up with a structure of seven principal smells: aromatic (warm cloves), fragrant (floral), ambrosial (musk-like), alliaceous (garlic-like), hircine (goat-like), foul (repulsive) and nauseating (disgusting).[14] But everyday aromas like fish, olive oil and bread didn't fit into his model and, although he never conceded it himself, it simply didn't work.

The Nose-Man

In the Age of Enlightenment, we can spy the beginnings of the ocularcentrism that would dog us in the future, through the Industrial Revolution and all the way up to the present day. Scientists, politicians and industrialists were now supposed to study and dominate the world primarily through their attentive gaze. This, along with the sense of hearing, would underpin civilized debate, intellectual activity and the hard sciences. According to notables such as James Cook and Alexander von Humboldt, an interest in other sensory experiences could, at worst, lead to insanity and bestial behaviour. On their travels back and forth across the continents, they had repeatedly observed that the more 'primitive' tribes were, the better developed their senses of smell and touch. And a society that smelled and touched a lot, it was said, did not do a great deal of thinking.[15]

The German biologist Lorenz Oken took matters even further, constructing a sensory hierarchy of the human races, topped by the European eye-man. The Asian ear-man came second, followed by the Native American nose-man, the Australian tongue-man and, in last place, the African skin-man.[16]

Both Linnaeus and Kant were fascinated by the new race theories and they saw it as perfectly natural for other peoples to submit to white dominance.[17] For the Europeans, civilization and the sovereignty of the eye soon became two sides of the same coin. This would not just help justify slavery but inspire ruthless colonization of the world.

The Rose

For centuries, roses had been a symbol of visual and olfactory perfection. In ancient Rome, their strong, complex fragrances were prized both by worldly folk and mystics, who associated them with the goddess Venus; later, in the Christian church, they came to be linked to the Virgin Mary. Through the Middle Ages and the Renaissance, their scents had served as protection against the plague and a crucial component in the flourishing of the perfume industry.

During the era of the Enlightenment, this changed abruptly. Whereas roses had previously been cultivated alongside other plants, they were now bred as a monoculture in separate flower-beds, and their scent was no longer among their sought-after properties. The central aim was to improve their appearance through perfection of colour and form.[18] In a matter of generations, many of history's most legendary rose scents vanished and it has never been possible to revive them. The fact that we are now left with scent-free Veranda Brilliant, Luna Love and Sand Castle roses is almost a mockery of the true nature of the flower.

9

Soda and Sulphuric Acid: The Smell of Prosperity

In his 1836 short story 'The Nose', the Ukrainian writer Nikolai Gogol tells the tale of Major Kovalyov, who discovers one fine day that his nose is missing. 'Damn it! What kind of trick is this!' he cries, spitting on the floor. 'If only there were something to take its place, but there's nothing!'[1] The nose is soon discovered baked into the local barber's breakfast loaf – perfectly intact, complete with the familiar wart on the left-hand side. But when Kovalyov visits the doctor to have it put back on again, he gets a chilly reception: 'believe me you'll feel just as good as when you had a nose. Now, as far as the nose is concerned, put it in a jar of alcohol; better still, soak it in two tablespoonsful of sour vodka and warmed-up vinegar.'[2] In other words, Kovalyov's doctor was perfectly in tune with the spirit of the age. The sense of smell, and by extension the organ of smell itself, was something you could easily do without.

Only if you were an officer, a barber or an enlightenment philosopher, mind you: in many professions the situation was quite different. For miners and sewage workers, the idea of going into work without a nose to alert you to dangerous levels of suffocating sulphur gases would have been absolutely inconceivable. And in many branches of industry, a good sense of smell was a crucial means of guaranteeing the quality of the products.

Craft Workers and Artisans

Craft workers have existed for more than 5,000 years and during the Middle Ages they formed guilds where apprentices learned from their masters. The procedure was simple: demonstration, observation and imitation – and a generous helping of secrecy. The initiates built on their knowledge through their own experience, without using measuring instruments, tables or academic science.

The quarrymen who excavated the materials for the new cities used their entire sensory apparatus. They judged whether the stone was suitable for interior or exterior use, in foundations, walls or roofing, based on the sound of the hammer blows, the creaking of the stone under pressure, how much it adhered to the tongue and how it smelt when they breathed on it.

And it was the same in the brickworks: here the workers had no access to thermometers or hygrometers, and had to rely on their own experiences of colour, consistency and smell to decide when the furnace was hot enough and when the clay product had been fired properly.

Work in the timber industry demanded insight into the different types of wood and their properties. All of them had their own olfactory signature – from the resinous aromas of split pine and spruce, which increase the further you penetrate into the solid heartwood, the fruity smells of pear and apple wood, the nutty scent of walnut and the malty smell of yew, all the way to the subtly meaty stink of larches, and oak trees that have acidified in bogs into material for window frames and tobacco pipes.

In beer-brewing, a good nose was a fundamental requirement for new recruits – to ensure that they could, for example, determine the quality of the malt. The worst malt smelt of rotten apples and the best of cucumber.[3] And when the returned barrels were cleaned, it was absolutely vital to get rid of the 'stinkers', which would ruin the brew in an instant.[4]

A good sense of smell was important in the rest of the food and beverages industry too.[5] This was particularly true when it came to the quality control of goods from the colonies, such as coffee, tea and tobacco, which were exposed to precise doses of sun and humidity to bring out the best aromas. Even Kant, who had completely dismissed the value of the sense of smell, was more reliant on skilled noses than he guessed when he started each morning with a period of meditation over a tobacco pipe and a cup of weak tea.

Arrival of the Engineers

Immediate sensory experiences contributed to optimizing the products, then. And as long as the customers were happy, the craftspeople themselves were left to plan and manage the production processes from start to finish. The work was not only varied but gave them responsibility.

This worker-driven form of production was universal in many industries right up until our own era. The steel industry was among the first to cave in after the American factory owner Cyrus McCormick II made it his life's work to 'weed out the bad elements' in the unions that had gradually replaced the guild system.[6] He invested in more machinery, employed engineers and dismissed much of the workforce. Although production fell, he had soon achieved his goal. He and the engineers took full control of production.

And that was the start of the form of production that has since become the established norm, controlled by the engineers. The sensory apparatus was now replaced by machinery, measuring instruments and automation.

From the last half of the eighteenth century, Europe underwent an industrial revolution. New enterprises sprung up everywhere, delivering goods to a rapidly growing urban population, and to rural communities where sickles, scythes and simple wheeled ploughs were replaced by mechanical reapers and ploughs with multiple razor-sharp shares. In addition to ironworks, textile mills, glassworks and paper mills were built, as were glue, margarine and cement factories. They used raw materials such as lime, coal, petroleum, fish oil, quartz, gypsum, timber and bone, as well as myriad new chemicals. Sulphuric acid was among the most important, used as a bleaching agent in the textile industry and a solvent in metallurgy, while soda played an essential role in the production of glass, soap, paper and dye pigments.

Ever more extensive stretches of the European landscape were invaded by noise – steady, rhythmic noise: 'hammer rings incessantly on anvil; there is an unending clang of metal, an unceasing clank of engines, flame rustles, water hisses, steam roars.'[7] Soil, water and air pollution rose to dramatic levels, and the most highly industrialized regions were blanketed in a grey haze. It blackened the buildings, stung the eyes and caused chronic bronchitis and other respiratory diseases.

And it smelt, in part owing to the use of coal fuel, while the production processes in the cement factories and gypsum works added an acrid, oniony stench of sulphur dioxide. But nothing could outstink soda production with its reek of gastric acid and rotten eggs.

> The gas from these manufactories is of such a deleterious nature as to blight everything within its influence, and is alike baneful to health and property. The herbage of the fields in their vicinity is scorched, the gardens neither yield fruit nor vegetables; many

INDUSTRY

ODORANT	GROSS FORMULA	MOLECULAR MASS [U]	BOILING POINT [°C]	ODOUR DETECTION THRESHOLD [PPB]
Hydrogen chloride	HCl	36	−85	275–1,000
Calcium sulphide	CaS	72	*c.* 1,600	0.5–300

A procedure known as the Leblanc process is used to produce soda from salt, sulphuric acid, limestone and coal, and the chemical reactions lead to massive emissions of hydrogen chloride and calcium sulphide. Both substances have low molecular weights and spread rapidly in the environment. Whereas calcium sulphide smells of rotten eggs with a hint of the sewer, the smell of hydrogen chloride is sharp and irritating, with a distinct element of chlorine. Since hydrogen chloride is also naturally present in our gastric juices as a bactericide, acid reflux offers us some indication of the aroma.

The Leblanc process was soon replaced with less-polluting production methods, but if you live in the vicinity of a cellulose factory, you will experience emissions of calcium sulphide from the bleaching processes. And the plastics industry continues to produce hydrogen chloride, especially in the manufacture of polyvinyl chloride or PVC – among the most widely used plastics. Admittedly, much of it is recovered during the production processes, so the emissions are relatively modest.

Like visible smoke, most odorous emissions have now been eliminated in European industrial production. This was probably aided by the fact that their effects are tangible, bearing in mind that there has been much less of a reduction in dangerous pollutants that can be neither seen nor smelt.

> flourishing trees have lately become rotten naked sticks. Cattle and poultry droop and pine away. It tarnishes the furniture in our houses, and when we are exposed to it, which is of frequent occurrence, we are afflicted with coughs and pains in the head.[8]

Security and Prosperity

Although many complained, the industrial emissions of the nineteenth century were usually regarded as unproblematic and even, at best, as cleansing and health-giving.[9] According to the experts, this was especially true in the chemical industry:

> When one goes into factories producing simple and oxygenized sulfuric, nitric and hydrochloric acid, for example, one is immediately struck by the odour of these acids, while the workers barely notice them and are disturbed by them only when they unthinkingly breathe in a great deal at one time.[10]

When the well-established German author Wilhelm Raabe tried to interest publishers in a novel dealing with industrial pollution in the 1880s, the manuscript was sent straight back to him. It would later go on to become a bestseller, but given the excessive economic and technological enthusiasm of the day, Raabe was seen as a killjoy. The time for highlighting the dark side of industrialization had not yet come.[11] Pollution and chemical stinks were, in practice, excluded from art and culture alike, although the contemporaneous paintings of naturalists like Gustave Courbet and Christian Krohg offer a suggestion of the stench.[12]

This well-disposed attitude has largely persisted. The author Margit Walsø speaks of her childhood in the aluminium-producing town of

Sunndal in the 1970s, where the fluoride emissions from the long electrolysis halls had a sharp and dusty metallic smell that the inhabitants perceived as acceptable, almost pleasant. It was associated with security, growth and prosperity, and for the teenage Walsø, it also carried feelings of 'anticipation, youth and freedom'.[13] She admits that, even in adulthood, the smell can revive those optimistic emotions, even though the fluoride emissions were harmful to people's health, made the conifers 'grow withered and yellow, and destroyed the teeth of the grazing sheep'.[14]

Our ability to readily accept smells that are associated with well-being and security is amply documented, especially when it comes to key firms in small industrial communities.[15] The same is true of industrial agriculture. Large-scale pig breeders and poultry producers spread acrid, sulphurous compounds in their vicinity yet studies show that the farmers mostly accept and are content with the smell.[16] Other residents tend to dislike it, however, and often become sick. In other words, the way we experience smell depends on our own personal stake in it.

Stockfish

The most serious industrial pollution tends to be linked to the engineer-managed form of production, with its intensive use of chemicals and constant introduction of new substances whose environmental impact is startlingly unclear. After taking over the ironworks in the late nineteenth century, the engineers soon took control of the production of paint, paper, bricks and bakery products. In the twenty-first century, only a handful of niches in the food and beverage industry still use sense-based working methods. There are, as yet, no engineer-managed alternatives in the production of exclusive cheese and wine, although we suspect that the use of AI may soon change that.

For now, the quality of stockfish (the Norwegian cured fish that is split and dried in the open air) is also established through full engagement of the sensory apparatus – and the product is divided into prima (first class), sekunda (second class) and a third category that goes by the name of Afrika. The nose is an utterly crucial part of the process. In this secretive line of work, which has remained unchanged since the Middle Ages, the professional sorter methodically sniffs the aroma of the belly in a series of short breaths. The smell 'must be healthy', the sorter declares, but he gives nothing else away.[17]

10

The Sweat of the Earth: The Smell of Fear

Towards the end of the eighteenth century, the French physician Jean-Noël Hallé took the Enlightenment philosophers to task over their disdain for the sense of smell. He could accept that it was a primitive and animal sense, but it was nonetheless absolutely crucial for our own self-preservation.[1] Like a kind of vanguard of the sense of taste, it could forewarn us of hidden dangers in our surroundings and, in many cases, the smells themselves could be malignant, he claimed. Hallé himself had experienced breathlessness and swelling of the tongue when he and a colleague had inspected sewage discharge along the highly polluted Seine on its passage through Paris.[2] And he suffered similar symptoms during his inspection of the city's hospitals with their musty stench of the sick and dying – 'the odor of decomposition . . . a mixture of the acidic, the sickly, and the fetid'.[3]

Decomposition

Hallé's theories could be confirmed through experiments. Chemists had diligently set to work gathering different smells to determine how they affected rats, dogs and rabbits. Although many of the results bore signs of haste and were plagued by unclear terminology and classification, there was little doubt that the scariest smells originated in nature itself, largely in the processes of biological decomposition.[4] From a broader perspective, chemists saw the Earth as being in a continual process of fermentation, while an invisible vapour – a kind of all-enveloping transpiration – rose up from the soil. And its extent was measurable, they claimed. A little over 4 litres of this earthly sweat was deposited annually on every single square foot of the Earth's surface.[5]

The resultant dangers were confirmed by the generally poor health of miners who spent their working days deep in the underworld itself. In fact all and any kinds of openings in the surface of the earth were associated with risk. Farmers were supposed to stay away from freshly ploughed fields, and farmworkers definitely shouldn't sleep in the furrows.[6] And, like an almost 2,000-year-old echo of Rome's Vitrivius, waterlogged soil, bogs and swamps were said to be especially bad. The very worst, though, was soil tainted by excrement and rotting corpses.

Miasma

In the early part of the nineteenth century, the noxious emanations were categorized under the term 'miasma', which comes from the Greek word for pollution. According to a popular manual written by the German physician Carl Bock – *Volksgesundheitslehrer* (Public Health Manual) – they could be sorted into the following categories: marsh miasma from bogs, earthbound miasma from graveyards, animal emanations from sewers and places where many people gather, and, finally, air miasma, which is borne on the wind and is of unknown origin.[7]

Anyone who came into contact with the miasmas risked dramatic consequences, which were quite similar to the effects of the devil's breath during the plague epidemics some centuries earlier. Now the miasmas were blamed whenever the cities of Europe and America were haunted by ever more frequent outbreaks of typhus and cholera, followed by waves of leprosy, measles, tuberculosis and above all malaria – whose name derives from the Italian *mala aria* or bad air. People were strongly advised to avoid places with miasmic air, especially in the evening and in foggy weather, when the risk was considered to be at its peak.

The Experts

The miasma theory had the unanimous backing of health experts, including the pioneering nurse Florence Nightingale, who believed it provided a 'satisfactory explanation for pestilence'.[8] She was convinced in particular that the white plague (tuberculosis) was triggered by unpleasant smells in the sufferers' homes. There was also little doubt about the link between malaria and the stench of boggy areas, even though no one knew anything back then about the malaria mosquito's connection to the problem. The fact that extensive drainage works appeared to have helped was considered good enough proof.

The medical experts were led by a German doctor, Max von Pettenkofer, professor of medicinal chemistry at the University of Munich. He was a true Enlightenment man who claimed that everything must be science-based, founded on chemical experiments and objective methods of measurement. He developed a number of special measuring instruments to determine the degree of ventilation in rooms, the humidity of indoor air and the level of carbon dioxide – the gas we humans exhale.[9] When the cholera epidemic hit Munich in 1854, Pettenkofer set to work with exceptional

SWAMP

ODORANT	GROSS FORMULA	MOLECULAR MASS [U]	BOILING POINT [°C]	ODOUR DETECTION THRESHOLD [PPB]
Ammonia	NH_3	17	−33	40–50,000
Methanethiol	CH_4S	48	6	0.0002–40
Hydrogen sulphide	H_2S	34	60	1–130

When you wade through a bog, the suction as you lift your boots will pump out the sweetish-sour smell that is most closely associated with miasma. It is a cool, intense odour that doesn't just contain various organic acids and a slight tang of ammonia, but consists primarily of various acrid sulphur compounds. Although hydrogen sulphide is the most dominant of these, there can also be large doses of methanethiol.

We recognize methanethiol from eccrine sweat, bad breath and intestinal gases – where it adds a rancid waft of rotten cabbage, often with a chemical character and a hint of fermentation.

Hydrogen sulphide, which is also present in intestinal gases, is created when organic materials from the plant and animal kingdoms decompose in the absence of oxygen; this is a substance we also find in the vapour from composters, in sewers, and in fermenting seaweed and

zeal, and his systematic investigations of the entire city resulted in what was known as the groundwater theory. Put briefly, the idea was that when urine and excrement rotted and fermented in the ground, they flowed into the groundwater, creating a gas that invaded the houses. That was why people whose homes were built on soil were hardest hit. If your house was built on rocky ground, you were safe.

kelp on the beaches. But what most of us think of is rotten eggs. Egg whites contain high levels of sulphurous amino acids, and hydrogen sulphide is created when they decompose. At the same time, though, other products of decomposition mingle with it, adding a whiff of cheese and urine. If the egg is boiled instead, you will get a cleaner smell.[10] In that case, the substance is created through what is known as a thermic reaction, and the longer the egg is allowed to boil, the stronger the smell will be when you crack it open. But it's only a matter of seconds before the hydrogen sulphide reacts again, creating sulphur compounds with a less pungent smell.

Even extremely low levels of hydrogen sulphide are detectable. However, the smell vanishes entirely in high concentrations, as if our sense of smell is paralysed. But this happens only in very unusual circumstances – in airtight manure cellars or large sewage works, or in the event of massive emissions from industries processing sulphurous oil and gas. That said, the effects on our health can be serious: paralysis, shortness of breath and even death. During the miasma hysteria of the nineteenth century, the amounts people inhaled were often far lower, and rarely reached levels that posed a health hazard, although they would probably have noticed some irritation of the eyes and mucous membranes in places with a high density of waste piles and latrines.

The Cities Grow

The cities had expanded dramatically over the course of the Industrial Revolution. Apart from sporadic sirens that marked the rhythms of the working day and the steady background roar of the factory machines, the streets were dominated by the clatter of horseshoes and iron-rimmed cartwheels against the cobblestones, along with clucking chickens, pigs in the backyards and herds of cattle on their way to the slaughterhouses. The differences between Roman and mediaeval cities were few. And apart from changes in the composition of the industrial odours, the streets were still filled with pungent mixtures of horse dung and chicken manure, ash and leather, smoke from the ovens and damp straw; of roasted chestnuts, eggs and freshly baked bread; of spices and herbs, and – not least – of unwashed people, because physical cleanliness was still the exception rather than the rule.

Halfway through the nineteenth century, though, waste and sewage were still the most feared odours. Here, too, conditions were unchanged since Roman times. Although the cities had spread out, little or nothing had been done to deal with the drainage problems. They had merely increased. In the case of faeces, experts like Jean-Noël Hallé applied four danger levels. First, the smell of fresh faeces, which wasn't considered especially harmful. Second, the smell of latrines. Third, the smell of pits. And finally, and most dangerously, the stench of cesspit clearing.[11]

Graveyards were also considered sources of miasma. They had grown slowly and steadily in tandem with the city populations and were bursting at the seams. There was much talk of slapdash procedures, of corpses stacked on top of each other for days before being buried – always in shallow graves.[12] On hot summer days, the stink of death and decomposition could seem ubiquitous.

Waste water and urine flowed through the narrow streets, along with blood from the butchers, and all of this combined to form a black fermenting mass beneath the cobbles. The puddles that formed helped spread pathogens when cartwheels sprayed walls and pedestrians alike.[13] In its autumn 1847 edition, the satirical magazine *Punch* made fun of the grim atmosphere that hung over London:

A glimmering vapour creeps over the ground,
You may see whence it issues – yon newly-raised mound;
Mark what spectre ascends in that horrible light –
Lo, the Vampyre Infection is rising to-night![14]

Miasma in the Home

The air inside the houses smelt of unwashed people, damp clothes, wood-fired stoves and especially the reek of the light sources, which were fuelled with fish oil and unpurified tallow fat. These smells were accompanied by the miasmas that seeped in through the outer walls of porous brick and lime mortar.[15] The closer to the ground you were, the worse, and upper storeys soon became more popular – although if the house was in the vicinity of a cemetery, the move did little good.[16]

Once the miasmas had invaded the building, they could impregnate the materials and persist for years. Even solid timber could be saturated all the way into the heartwood. After old pus was removed from the walls and ceilings of a hospital, the stink was so intense that it 'even surpassed the odor of a dissection room'.[17]

Confirmation of the ravages of the miasmas came from all over Europe, including Norway. On the ground floor of a small apartment building in Enerhaugen, Kristiania, two families of four adults, seven children and two pigs had been living in a room that served as kitchen, living room and bedroom. When the cellar, which had been filled with stinking water, was pumped out, they found cat carcasses and enormous numbers of rotting mice and rats. All of the residents had contracted cholera and only one of them survived.[18]

The Slum in Bangkok

In the early 1990s, the Israeli social anthropologist Erik Cohen spent a few months in a slum building on the outskirts of Thailand's megalopolis capital, Bangkok.[19] Its residents were poor people who had moved into the city from the countryside and the site had once been agricultural land where there were rice paddies and orchards. On either side of narrow plank-paved alleyways were houses built on stilts that backed onto open drainage channels. All the waste was piled up between the houses, or dumped into the channels along with the sewage. Eventually, this created simmering black mud that would swirl up and flood the entire district in rainy season.

The stench must have been like a kind of mega-miasma, and it was particularly intense at night, when it mingled with the exhaust of hundreds of mopeds and the odour of busy food stalls that cooked with charcoal. It soon seeped into Cohen's room through the gaps between the floorboards and he felt a faint nausea, which grew worse as the night drew on.

But although these conditions bore clear similarities to those at the workers' home in Enerhaugen a hundred years earlier – rat density was probably even higher – the residents seemed indifferent, much less bothered. They did little or nothing to combat the stench that was continuously pumped out by their environment.

Cohen assumed that this striking lack of interest may have been due to the population's rural origins, even though it was ages since they had left, because in the agricultural communities, waste and sewage were part of a positive ecological cycle and were used as life-giving fertilizer. The stink roused no problematic associations. Quite the contrary.

In his 1862 novel *Les Misérables*, Victor Hugo touched on this subject:

> The garbage heaped up beside the stone blocks, the tumbrils of mire jolting through the streets at night, the awful scavengers' carts, the fetid streams of subterranean slime that the pavement hides from you, do you know what all this is? It is the flowering meadow, it is the green grass, it is the marjoram and thyme and sage, it is cattle, it is the satisfied lowing of huge oxen in the evening, it is perfumed hay, it is golden wheat, it is bread on your table, it is warm blood in your veins, it is health, it is joy, it is life.[20]

This may help explain why the fear of miasma was a largely urban phenomenon stemming primarily from the city burghers.

11

The Cast-Iron Plant: The Smell of Class

In 1820 the British aristocrat Arthur de Capell Brooke travelled the length of Scandinavia, from Gothenburg in the south to Finnmark in the north. He published an account of his travels, which one reviewer described as 'written with the feelings and in the style of a gentleman'.[1] Brooke wrote enthusiastically about the wild bare mountains and fertile valleys, but also mentioned some of the more dismal experiences of his journey, including a visit to Breiden Farm in Gudbrandsdalen. A well-to-do farmer – 'stout, corpulent, and greasy' – lived here with his seventeen children, aged relatives and farm hands in thoroughly deplorable conditions: 'the dirt was inconceivable; and I anxiously looked forward to . . . be[ing] released from it.'[2] When, by accident, he opened the door to a kind of dormitory, he witnessed a scene that was not just repulsive but astonishingly bizarre:

> In five or six large beds, or rather wooden cribs, near twenty persons of both sexes, perfectly naked, were lying together in heaps; and the dark copper-coloured skins of some, contrasting with the whiteness of others, rendered the group still more extraordinary. To complete it, on the ground, several large pigs were enjoying the sweets of repose, and responding

> with drowsy grunts to the snores of, I might almost say, their fellow swine.[3]

The stench of boar, bad breath, sour sweat and flatulence must have been overpowering, but that will have come as no great surprise to Brooke: it was a well-known fact that the lower classes stank, and much of the European bourgeoisie was convinced that this was also a health hazard.[4]

Miasmatic Ovens

As the Industrial Revolution progressed, the class divisions in the cities became increasingly stark, and the working-class districts, in particular, were clearly demarcated. Yet even a short stroll might prove risky, especially if your route took in dubious restaurants and other establishments where all kinds of people still gathered – generally referred to as 'miasmatic ovens'.[5] And people questioned whether the lower classes should even be given access to the many museums that were now being established in the big cities. Some claimed that their stench was not only a danger to public health but could pose a threat to the works of art.[6]

Even churches weren't free of such prejudice. Mark Twain reports the complaints of a frustrated churchgoer who could detect the stink of a worker who had just entered the church from his own front-row pew: 'If you are going to kill the church thus with bad smells, I will have nothing to do with this work of evangelization.'[7]

To avoid these insalubrious emanations, the bourgeoisie once again retreated into nature. But they steered clear of the villages where 'all seem to be a part of the same slime. Where all the women bawl, all the children weep, and all the men sweat.'[8] Instead they withdrew to the highlands, into the mountains and out onto the open moors, which eventually became the sites of countless health resorts offering air cures and air baths. Or they went to seaside resorts because the sea air was considered particularly pure.

Hygiene

In Ibsen's 1882 play *An Enemy of the People*, the social critic Dr Stockmann leaves his listeners in no doubt about the gravity of the situation:

> In a house which does not get aired and swept every day – my wife Katherine maintains that the floor ought to be scrubbed as well, but that is a debatable question – in such a house, let me tell you, people will lose within two or three years the power of thinking or acting in a moral manner. Lack of oxygen weakens the conscience.[9]

Hygiene and cleanliness were now on the bourgeois agenda. And this didn't just apply to the home: the body was a target too. With advances in medical science, the Renaissance fear of bathing had been dismissed, along with the theory that open pores made the body more vulnerable to disease. On the contrary, it was now claimed, the risk of illness was far greater when sweat and other corrupted bodily fluids were sealed inside.[10] This had been documented in various ways, including macabre experiments in which animals that were coated in tar quickly fell sick and died when their skin was unable to breathe.[11]

Face, hands and feet were to be cleaned with particular frequency, but the rest of the body was also subjected to regular routines.[12] The

MONEY

ODORANT	GROSS FORMULA	MOLECULAR MASS [U]	BOILING POINT [°C]	ODOUR DETECTION THRESHOLD [PPB]
1-octen-3-one	$C_8H_{14}O$	126	59	0.2

There was one odour even the bourgeoisie couldn't escape: the smell of money. The idea that money stinks isn't just a much-used metaphor for economic corruption. Money does in fact have its very own smell.

Brand-new banknotes give off only a faint whiff of paper and printer's ink, but a brief period in circulation is all it takes for plastic and leather wallets, trouser pockets, cashier's tills and above all sweaty human hands to set their intense stamp upon the absorbent material. While the result is certainly interesting, it is impossible to separate out any typical odorant molecules.

Coins are a different matter altogether. True, we can't smell metals, so a highly polished coin smells of nothing at all. But when the sebum on our skin reacts with the surface of the coin, the ketone compound 1-octen-3-one forms – the same molecule that contributes to the smell of blood – and it makes little difference whether the coins are made of nickel, copper or zinc.

The odour threshold is low and the smell can be rusty, metallic and sweetish, with a hint of mould. Many will associate it with brass candlesticks, campfires and fireplaces. It is simple and slightly watery in nature, and occurs in doses that are hardly high enough to trigger the same agitation that the stench of blood provoked in Roman amphitheatres 2,000 years ago – although who knows?

new doctrine encompassed clothes and bed linen too, which were now expected to be changed much more often than before.

All this barely touched the working classes, whose cramped and poorly furnished homes offered little opportunity to follow such strictures. But there was no way out for anyone employed as a parlourmaid or servant in one of the finer households. In *Veiledningsbok for tjenestepiker og unge husmødre* (Book of Guidelines for Serving Girls and Young Housewives), Henrikka Høeghs Schmidt sums up the principles that would prevent servants from bothering the guests unduly: 'Guard against the smell of sweat and bad breath . . . or breathing so audibly that anyone would take you for a steam engine!'[13]

All Smell Is Disease

In 1709 a French parfumier decided to ensure that the different social classes were easier to identify by providing them with their own scents: a regal perfume for the aristocracy, a bourgeois perfume for the middle classes, while those at the bottom of the pile would have to make do with a simple disinfectant.[14]

Some way into the nineteenth century, this was no longer a relevant issue. Improved hygiene routines meant that there was less need for the copious use of perfume that had previously camouflaged body odours. This was especially true for women of the bourgeoisie, who now used floral scents of rose and jasmine at most, but preferably no perfume at all.

'All smell is disease,' declared outspoken English reformer Edwin Chadwick in 1846.[15] And it didn't matter whether the smell was good or bad. He was zealously dedicated to improving public health in Britain and was probably well meaning enough. We might almost suspect him of being the inspiration for 'Holger Drachmann', the poem by another equally ardent social critic, Bjørnstjerne Bjørnson:

Just this it is that we are needing now:
The spring, the spring! These stifling fumes we bear
Of royal incense and of monkish snuff,

Of corpses in romantic cloak and ruff,
Are bad for morals and for lungs: Fresh air![16]

Aspidistra

It was at this point that an Asian potted plant known as the aspidistra made its way into the parlours of the bourgeoisie. An evergreen plant, it was hardy and drought-resistant, but most important of all, it had no smell whatsoever – a feature that must surely have contributed to its nickname: the cast-iron plant. But while this may have made it almost a symbol of the triumph of ocularcentricity, its limp evergreen leaves and sparse flowers offer disappointingly little to please the eye.

In Great Britain, the aspidistra remained popular well into the twentieth century, eventually becoming the embodiment of 'dull middle class respectability', according to the 1933 edition of the *Oxford English Dictionary*. In his 1936 novel *Keep the Aspidistra Flying* this is precisely the angle British author George Orwell exploits. Driven by a fervent desire to escape the pointless quest for money and status that typifies his era, the protagonist Gordon tries in vain to kill his aspidistra: 'Many a time he had furtively attempted to kill it – starving it of water, grinding hot cigarette-ends against its stem, even mixing salt with its earth. But the beastly things are practically immortal.'[17]

Class Code

While the noblest senses, sight and hearing, are associated with intellectual work, taste, smell and touch are linked to manual labour. To take the most extreme variants: at one end, we have the unfettered artist with his exalted spirituality – art for the sake of the soul; and at the other, the basket-weaver's trivial, coarse handwork, which fulfils a practical social need.[18] Just like the relationship between male and female work, the ranking of intellectual and manual work – from architect to carpenter and engineer to industrial worker – has been a feature of all class-based societies since antiquity. In both cases, the point is to affirm and preserve the social pecking order. And all this came to a head when the massive industrialization of the nineteenth century led to the expansion of the working classes.

The bourgeoisie had by now become so thoroughly refined and intellectual that the contrast between them and the growing lower-class population had never been greater. As a result, smell became an even clearer class marker, ranging from the lofty absence of scent among the elites down to the slurry of foul stenches that confirmed poverty, uncouthness and inferiority.[19] And this has persisted into our own age.

The Stink of the Working Classes

Even George Orwell, ever the champion of the oppressed, was both horrified and nauseated after spending several months in a working-class slum in 1937 – the year after he mocked the bourgeois passion for aspidistras. But most of all, he was in despair:

> Here you come to the real secret of class distinctions in the West – the real reason why a European of bourgeois upbringing, even when he calls himself a Communist, cannot without a hard effort think of a working man as his equal. It is summed up in four frightful words which people nowadays are chary of uttering, but which were bandied about quite freely in my childhood. The words were: *The lower classes smell.* [. . .] And

> even 'lower-class' people whom you knew to be quite clean – servants for example – were faintly unappetizing. The smell of their sweat, the very texture of their skins, were mysteriously different from yours.[20]

According to Orwell, this stink was a serious obstacle to the achievement of a just world. People could deal with a rigid temperament, religious wrangles and differences in ethical standards – but not a disgusting smell. Perhaps it's hardly surprising that he soon isolated himself on a Scottish island.

Mark Twain, far less priggish, had nothing but scorn for the bourgeoisie's distaste, especially an arrogant churchgoer's whining about the sweaty smell of the American congregation:

> We have reason to believe that there will be labouring men in heaven; and also a number of negroes, and Esquimaux, and Terra del Fuegans, and Arabs, and a few Indians, and possibly even some Spaniards and Portuguese. All things are possible with God.[21]

And he declared that the grumbler would not have passed muster as one of Jesus's disciples, 'because he could not have stood the fishy smell of some of his comrades who came from around the Sea of Galilee'.[22]

12

Damp Cows and Shallow Graves: The Smell of Confusion

In the last half of the nineteenth century, it was important to separate the clean, healthy and living from the dirty, diseased and dead. Since the smell itself, the miasma, was thought to be the cause of disease, only the sense of smell could identify the source.

Super-sniffers were recruited in the big cities. All of them were men and they never touched tobacco and, if we are to believe fiction of the day, they worked according to established traditions. In Edgar Allan Poe's 'The Murders in the Rue Morgue', the sharp nose of the detective, C. Auguste Dupin, was a decisive factor in solving the case.[1] And Arthur Conan Doyle's Sherlock Holmes, who obviously didn't get the message about not smoking, declared that a skilled detective should be able to identify at least seventy-five different perfumes.[2]

Sanitation

Armed with these ideals, the miasma detectives set out on their urban explorations. And, as we know, the task they faced was often tricky. Every smell can consist of several different chemicals and our nose quickly becomes accustomed to it. To avoid following false trails, they had to take constant breaks in their hunt, which they conducted systematically – up and down the streets and alleys, possibly all the way out to the city's outskirts and beyond – until the source of the smell was located. Afterwards, the findings were summed up in an olfactory map, on which trouble spots were marked in different colours according to type and intensity.[3]

The ensuing clean-up operation could be a challenge, especially in the older, more labyrinthine city districts, which teemed with light industry, makeshift refuse solutions and random sewage systems. Many places were dealt with summarily: old town centres were demolished and replaced by clean, straight avenues lined with new buildings. These were far easier to air and not only satisfied the demands of hygiene but improved fire prevention. This was also a consideration when wooden and half-timbered houses were replaced with mineral-based constructions made of granite, sandstone, limestone and slate. And on streets and squares in the most distinguished districts, cobblestones were often replaced with solid paving stones to seal the almost bottomless Swiss cheese of miasma beneath the ground.

Sewage and Cemeteries

As part of the sanitation process, cemeteries were also moved to the outer edges of the cities. And whereas they often used to consist of shallow mass graves, individual and much deeper graves were now recommended, separated by a distance of 1.2 metres (4 ft) to ensure that 'the morbific rays' did not blend into 'deadly brews'.[4]

At the same time, sewage treatment was improved: first and foremost, outdoor privies were emptied more frequently, and their contents were then collected in sewage beds and supplemented with lime and bog soil to ease the worst of the stench. Sewerage systems were gradually expanded, and they were followed by water closets and utility sinks, the ultimate means of eliminating – or at least displacing – odours. But all this took time. Only 28 WCs were operating in Kristiania in 1864, all of them in the capital's wealthy west end.[5] And well into the twentieth century, the city was still divided in two, with WCs in the west and backyard privies in the east.

White Linen

But exterior spaces and communal functions weren't the only things that needed sanitizing: houses did too. Those who could afford it were advised to seal their cellar floors with asphalt – 'as impervious as glass' – and told that it was also best to use compact mineral materials when building.[6] Timber couldn't be entirely dispensed with, since it was needed to provide support in storey partitions and roofs. But since all biological material undergoes a continuous process of decomposition and decay, every joint and crack had to be assiduously sealed. The same went for porous brick walls, which should ideally be covered with glazed tiles placed edge to edge. If this was a bit too fancy for most, plaster and whitewash were reasonably good alternatives.

The white surfaces – especially if they were shiny and reflective – were almost like a visualization of the olfactory sterility.[7] And the impression was topped off by the white bed linen and white clothing of the residents.

The recommendations even applied to layout. Aroma-rich food preparation, which had previously taken place beside the living areas in houses, was now banished to a dedicated room some distance away. Where possible, special bathrooms were built even further out on the periphery to take care of the adults' hygiene needs, preferably

STONE

ODORANT	GROSS FORMULA	MOLECULAR MASS [U]	BOILING POINT [°C]	ODOUR DETECTION THRESHOLD [PPB]
Hydrogen sulphide	H_2S	34	60	1–130
Geosmin	$C_{12}H_{22}O$	182	270	0.4

While igneous rocks such as granite and gabbro are odour-free no matter how they're treated, sedimentary forms of slate, limestone, sandstone and flint may contain aromatic compounds that are released when the rocks are split, cut or hacked. When you crush a flintstone you may pick up wafts of menthol, latex, resin and other aromas that have lain stored in the rock ever since its formation during the hot, humid Cretaceous period more than 100 million years ago. And with a bit of imagination, you may picture echinoderms and plesiosaurs and endless rainforests burgeoning with epiphytes and the very first flowers.

Yet the dominant smell produced when you work with material like limestone originates from fragments of sulphurous minerals, which rapidly decompose further when exposed to humidity and oxygen. One of the decomposition products is hydrogen sulphide and, since the stench of swamps and rotten eggs was almost synonymous with miasma in the nineteenth century, it isn't unreasonable to imagine that many construction workers must have gone to work uneasy.

Once processed, however, all the commonest types of stone appear odourless, although only as long as they're dry. If you sniff their surfaces after rainfall, you'll pick up the characteristic smell of 'wet stone' – raw, fresh and slightly sweet, with a hint of wax, moss, cheese and boggy forest, and often a touch of fish and metal. It is composed of many different odorants, including earthy geosmin, and is known as petrichor, from the Greek words *petros* for stone and *ikhor* for blood of the gods. The smell doesn't originate from the stone itself but from the surface membrane created by the life of the city: exhaust fumes, industry, algae, fungi and fragments of plant, animal and human matter that have accumulated on the stone's surface over time.

with a separate cubicle for the WC. All this was complemented with a lavish assortment of window vents, ventilation channels and specially adapted heating systems designed to ensure a steady supply of fresh air.

Cleanliness

The countryside usually escaped the most drastic recommendations, but here too action was required. In western Norway, peripatetic teachers united to appeal for better ventilation and cleanliness. They refused to accept the working conditions in the peasants' cottages, which were cramped, filthy and nauseating. The children grew drowsy, they said, and they themselves suffered from headaches.[8]

The sociologist Eilert Sundt was sceptical about these complaints, citing the sound health and robustness of the local population. In his major 1869 study *On Cleanliness in Norway*, he went a step further, questioning the need for what he described as excessive hygiene. He dismissed, for example, the medical profession's view that there should be a clear division between the living quarters of people and livestock. Nor did he see anything wrong with the custom of young people sleeping alongside cows, pigs and poultry; indeed, he produced a statistic which demonstrated that the people who lived where this tradition was most widespread – such as Gudbrandsdalen, central Norway – also lived the longest, 'and must therefore enjoy the soundest health'.[9]

The Bacteria Theory

As early as 1854, the British epidemiologist John Snow had published a study of a cholera outbreak in Soho, London, in which he concluded that neither air nor smell was the cause of people's illness. The origin of the disease was the drinking water, which had been polluted by the excrement of the sick. In other words, cholera was linked to the water supply rather than poor cleanliness.

In 1873 the Norwegian physician Gerhard Henrik Armauer Hansen discovered the leprosy bacterium and, several years later, the German physician Robert Koch identified first the cholera bacterium and then the bacterium that causes tuberculosis. Soon it became clear that malaria was also caused by parasites and not miasma, although it would be some years before the role of the malaria mosquito was identified.

Koch's fellow-German Max von Pettenkofer, now an elderly man, remained an unwavering believer in the miasma theory. So unwavering that, in 1892, he drank a glass of fresh cholera bacteria culture to prove Koch wrong. When the only consequence was a mild bout of diarrhoea – probably because a case of cholera in 1854 had given him immunity – that merely confirmed him in his belief. But something must have ailed him all the same: soon afterwards, he descended into a prolonged depression and took his own life at the age of 83.[10]

Indoor Environment

By the turn of the century, physiologists had determined that the problems with the indoor environment – headaches, sluggishness and poor concentration – had more to do with heat and high humidity than uncleaned cottages and filthy schoolchildren.[11] The Scottish physician John Scott Haldane conducted an experiment involving an overcrowded room and two test subjects, each equipped with a snorkel.[12] One of them remained in the room but had his oxygen supplied from outside, while the other was placed outside in the open air but breathed in the stuffy indoor air, which was full of sweat and bad breath. The result left no room for doubt: the subject who had stayed indoors felt just as unwell as the other people in the room, whereas the person who stood outside was unaffected. The quality of the air the subjects were breathing had played no role.

Although, in principle, this finding let odour off the hook, it failed to significantly shift attitudes. Miasma theory was simply too firmly embedded in the culture. That must be why barely anyone raised an eyebrow when the mildly paranoid Swedish playwright August Strindberg claimed that a bout of illness he suffered in summer 1896 had been caused by Edvard Munch, who had tried to kill him by transmitting miasmatic gases through the wall.[13] And, as we shall see later, the dogma has survived into our own century.

The Warning Nose

Although the theory of a direct link between smell and infectious diseases had now been rejected, at least by bacteriologists, there was little doubt that smell was still important when it came to health. Based on our previous olfactory experiences, it can warn us of dangerous substances that may accompany certain odours. Spoiled meat and fish will, for example, emit the unmistakable, bittersweet stench of cadaverine and putrescine even though the actual causes of any food poisoning will be odourless bacteria such as salmonella, E. coli and botulinum. Similarly, the harmful substances in mould, known as mycotoxins, are rarely responsible for the otherwise characteristic smell of mushrooms.

Yet some odorants *are* harmful in themselves – in particular, certain volatile ingredients in industrial emissions, car exhausts and synthetic interior products, building materials and perfumes. In rare cases, natural odorants may also be harmful, such as ammonia and some sulphurous compounds, but the damage involves doses that rarely occur in the environment and – it hardly seems necessary to add – absolutely never in human secretions, even from the sickest of the sick.[14]

Social Reforms

Although it was baseless, the fear of miasmas did at least contribute to some constructive reforms. General levels of contagion fell after sick people were isolated from the healthy segment of the population. The draining of the swamps reduced the malaria epidemics, the construction of sewerage systems resulted in cleaner drinking water and the abandoned cemetery sites were redesigned as lush parks, places for play and outdoor activities.

Miasma theory also helped improve conditions for the working classes. By building public baths and allotments, and introducing measures to reduce overcrowding by way of larger, lighter and better-maintained apartments, the bourgeoisie did not just ensure that they would have access to a healthy, stable workforce. They also reduced their own risk of contagion.[15]

Without miasma theory, the cities wouldn't have been as well prepared as they were to tackle pollution from the motor traffic that invaded the streets just decades later; and buildings without extensive ventilation systems would not have been capable of dealing with the many harmful chemicals that invaded the construction sector towards the end of the twentieth century.

The Dark Side

In other words, an incorrect theory led to valuable countermeasures. A happy accident. But something was lost too. The new architecture was built according to standardized templates and retained few of the individual adaptations and vibrant styles that had been the hallmark of older buildings. Nowadays, that kind of architecture is found, if at all, in poor districts and remote rural communities that were never affected by the miasma-driven sanitation efforts.

At the same time, impermeable surface treatments, the covering of heavy constructions and a reduction in porous materials meant that new buildings were less able to regulate and stabilize their own room temperature and humidity. These jobs were therefore left to the ventilation system, which then became indispensable. And ever since then, ventilation systems have become more and more powerful.

13

Alabaster Brows: The Smell of Longing

In 1880 the French bourgeoisie were shocked by *Le Gousset* – 'The Armpit' – written by a 32-year-old Parisian flâneur called Joris-Karl Huysmans. First he describes the smell of a group of peasant women out strolling in the afternoon sun:

> It was excessive, terrible; it stung your nostrils like a flask of ammonia, or rather it gripped you, irritating the mucus membranes with a musky, gamey smell, like wild duck cooked in olives and the sharp odeur of shallots. But in truth, this emanation had nothing repugnant or vile about it; it harmonized with the strong smells of the countryside, like something expected; it was a pure note, the sultry cry of the human animal complementing the pungent melody of livestock and of woodland.[1]

And then he turns to the city women:

> an aroma of valerianate of ammonia and of urine is at times brutally accentuated . . . its range traverses the whole keyboard of the sense of smell, sometimes hinting at the insistent scents of seringia and elder-flower, sometimes recalling the soft perfume of one's fingers after having rolled and smoked a cigarette

> ... Never were women more desirable than at these moments, with their Oxford wool dresses moulding their bodies from top to toe, clinging to them like the damp chemises that imprison them underneath.[2]

Baudelaire

Bold words, these, from the young author: they challenged not only the refined feelings of the bourgeoisie but the dogma of miasma (bacteria theory had not yet been introduced and smell was still viewed as unhealthy and terrifying). Huysmans wasn't the first writer to plough this particular furrow, though. Back in 1856 the French poet Charles Baudelaire had published his notorious poetry collection *Les Fleurs du mal*, including the poem 'Léthé', which contained the following lines:

> I want to hide the throbbing of my head
> In your perfume, under those petticoats
> And breathe the musky scent of our old love
> The fading fragrance of the dying rose.[3]

This was strong stuff – far too strong. Baudelaire was hauled before the court where the state prosecutor Ernest Pinard called for the poem to be banned: it must never reach the ears of the general public – or their noses, for that matter.

There was no court case for Huysmans, though. Since Baudelaire's escapades, a robust counterculture had evolved, and its branches penetrated deep into the bourgeoisie.

The Decadents

The late nineteenth century was a time of prosperity, and faith in the future was high; yet many artists and intellectuals did not share the optimism of the age. They coined the term *fin de siècle*, which captured a certain melancholic sense of a waning era. Oddly enough, the most fervent of them referred to themselves as decadents, but this was a conscious provocation. The Decadents' intention was not to appear decadent themselves, but rather to unmask the decadence they thought they saw in their own age. At the same time, they claimed the right to pursue their artistic development free of outmoded authoritarian norms. In poetry's 'supernatural world of beauty', Baudelaire had declared 'desire pure', 'melancholy amusing' and 'despair noble'.[4]

The point was to relinquish control, and abandon oneself to one's emotions and sensual experiences. Art was not just meant to open eyes and ears; it should also penetrate the nose and lungs. And this was the very core of the rebellion against the constricting conventions of the bourgeoisie. Literature, which had been odourless throughout the century, suddenly took on a pungent stench.

In 1884, the year after the discovery of the cholera bacterium, Huysmans followed up his scandalous armpit sketch with *À rebours* (Against Nature) widely regarded as a kind of manifesto of the Decadent movement. In the novel, we meet a thirty-year-old aristocrat, Jean des Esseintes: frail and pallid, with cold eyes, hollow cheeks and dry hands. He is on the verge of suicide, having tired of the luxurious Parisian life he has pursued in 'a craving to sully what memories he retained of his family with acts of sensual depravity, a furious desire to expend his lustful frenzy on cushions of soft flesh and to drain the cup of sensuality to its last and bitterest dregs'.[5] After losing all interest in science, economics and politics too, there is only one way out. He will seek refuge far from civilization and explore the depths of his own sensual world.

He settles in a rural villa near Fontenay-aux-Roses that is equipped with an extensive laboratory. In order to ensure perfect silence, he has all the hinges oiled, all the floors padded with thick carpets, and provides the old couple who are his only servants with felt slippers.

Experiments

Then he sets to work following strictly methodical principles. First he fills a bathtub with water, adds some sodium chloride – cooking salt – together with a spot of magnesium and calcium, as found in mussels and crustaceans. After that he takes out a 'twist of rope bought for the occasion from one of those enormous roperies whose warehouses and cellars reek with the smell of the sea and sea-ports'.[6] Finally, he looks at a photograph of a seaside casino before reciting the text from a guidebook that depicts a beautiful bathing resort. All at once, he finds himself beside the sea, experiencing 'an illusion of sea-bathing which will be undeniable, convincing, and complete' and senses that he can 'so substitute the vision of a reality for the reality itself'.

And thus he continues, one experiment after another, and all of them combine to form the plot of the book. In one such experiment he showers the room with extracts of assorted flowers to create a baseline of a flower meadow. Then he adds carefully calibrated amounts of tuberose, orange and almond blossom, 'and immediately artificial lilacs came into being, while linden trees swayed in the wind'. To this background he adds 'a light rain of essences that were half-human, half-feline, smacking of the petticoats . . . the natural appearance of laughing, sweating, rollicking pleasures out in the sun'. Next he fans the air gently while warming a pellet of styrax resin between his fingers.[7] 'Then as if by magic, the horizon was filled with factories, whose fearsome chimneys belched fire and flame like so many bowls of punch.' When he put away the resin, the factories vanished instantly. And after he had mixed a few drops of 'New-Mown Hay' perfume into the restored aroma of meadows and lindens, 'there rose piles of hay, bringing a new season with them, spreading summer about them in these delicate emanations.'[8]

The experiments become increasingly extreme and in the end, Esseintes mixes all the essences he possesses into a single blend and 'the suffocating room was suddenly filled with an insanely sublimated vegetation.' It all gets too intense. He hurls himself towards the window and opens it wide, but finds no relief. Instead, he feels that the air is 'assailing his jaded nostrils, shaking anew his shattered nerves and throwing him into such a state of prostration that he fell fainting, almost dying, across the window-sill'. It is all too much for the exhausted aristocrat, who returns to Paris as soon as he possibly can to resume his old life: 'Ah, but my courage fails me.'[9]

And thus Huysmans suggests a sensual world that is both beautiful and magnificent but also as treacherous as a kingdom of the dead or a crazed acid trip.

The Giant Tortoise

So far, all this remained in the realm of literature and thought experiments. Because, on the whole, the Decadents were all talk and no action. Their own lives were thoroughly in step with bourgeois ideals: neat in dress, cultivated in speech and not given to practical antics of any significance – with the exception of the occasional eccentric. Comte Robert de Montesquiou-Fézensac was undoubtedly one such man. The main inspiration for Huysmans' book, he had previously tried his luck as a writer with *Pays des aromates* (Land of Scents) but had enjoyed little success.[10] It was as a practitioner that he would be remembered. During a soirée at his palatial townhouse, he served refreshing 'charismatic' drinks to his guests as they were sprinkled with meticulously selected perfumes. The only glint of light came from the jewel-encrusted shell of a giant tortoise that crawled disconsolately around on a Persian rug, glittering with genuine diamonds, sapphires and amethysts.[11] It's like an echo of those feasts for the senses that the Roman hedonists used to throw.

Alabaster Brows

The Decadents' fundamentally sensual approach to existence soon infected the bourgeoisie, whose lives were otherwise spent indoors with odour-free servants and aspidistra plants. In the summer months, though, they set off for mountain and coast, to sanatoria, spa hotels and seaside resorts – clad in loose-fitting clothes, bonnets or broad-brimmed hats – to cleanse themselves in the fresh air.

The well-to-do patrician merchant family in Thomas Mann's *Buddenbrooks* travels to the sandy beaches of the Baltic coast but only for brief stays.[12] The entire family is hypersensitive to sea mist and, with their nervy and refined dispositions, thin fingers, sharpened nails and alabaster brows, they appear to be creatures of pure spirit. In this sense, they satisfy the decadent ideal of the *fin de siècle*: women faint at the drop of a hat, and melancholy and shattered nerves are almost proof of spiritual power and class.[13]

Weary Men

The sense-based quest for authenticity and truth spreads almost by osmosis through European culture.[14] We soon find it again in all art forms, from music to architecture and the visual arts, from German impressionism to the British Arts and Crafts movement. Yet it still plays out most frequently in literature.

In Arne Garborg's *Trætte Mænd* (Weary Men), the hero Gabriel Gram is paralysed by the neuroses and melancholy of his age, while Oscar Wilde's *Picture of Dorian Gray* is haunted by a man who is entirely without a soul.[15] Like Esseintes in Huysmans' book, Wilde's main character is directly inspired by Montesquiou-Fézensac. The duke is also a close companion of Marcel Proust, who will soon give the world that memory-laden scent of linden tea and madeleines in *In Search of Lost Time*.[16] But Ireland's James Joyce was the first writer to get to the very heart of the matter in his masterwork, *Ulysses*.[17]

Here, he unabashedly poses questions about whether men and women smell different. Whether the hair of passionate women has a more powerful scent. Whether rich bodies smell different from poor ones. Why torn-off toenails have such a fascinating smell. Whether a cellar smells stronger if it is dark. How a fart spreads in a room and how long it lingers there. And where do these bodily odours come from anyway – from food, nether regions, armpit or neck? And last but not least: can the nose override the brain?

All these questions are left hanging because science still knows far too little about why things smell: is it material or pure spirit?

The Soul's Aroma

Many of the Decadents, including Huysmans and Garborg, were fascinated by Spiritualism, an exotic movement launched by the Russian occultist Madame Blavatsky in the 1870s.[18] With the help of inspired mediums, contact could be established with the dead through knocking, moving furniture and invisible bells. And when the soul eventually materialized in the form of ectoplasm – a milky white slime that bubbled out of the medium's mouth – it was accompanied by a thoroughly distinctive smell. According to enthusiasts including crime writer Arthur Conan Doyle, it had an unmistakable smell of ozone.[19]

OZONE

ODORANT	GROSS FORMULA	MOLECULAR MASS [U]	BOILING POINT [°C]	ODOUR DETECTION THRESHOLD [PPB]
Ozone	O_3	48	−112	7–30

The ozone layer in the stratosphere protects us against intensive doses of ultraviolet radiation from the Sun. Prior to thunderstorms and tempests, small amounts of ozone are drawn down to the surface of the Earth and spice up the familiar smell of summer rain.

Ozone consists of three oxygen atoms, whereas the more widespread oxygen molecule has only two. The substance forms naturally when oxygen molecules in the air split into free oxygen atoms, which then react with other oxygen molecules to become ozone. The split itself happens via photochemical reactions that are often triggered by ultraviolet light. And the process is stimulated by contact with organic compounds and nitrogen. That is why copious amounts of ozone are produced in heavily polluted areas on sunny days, especially in the presence of nitrogen emissions from motor traffic. But direct concentrations increase dramatically during electrical discharges in stormy weather, around short-circuiting power plants or in the vicinity of old photocopying machines. Given its low molecular mass and low boiling point, the substance disperses rapidly into the environment.

Whereas ordinary oxygen is odourless, ozone emits a piercing, slightly dry and sweetish metallic smell with a touch of chlorine. The odour detection threshold is low but may vary greatly from one person to the next. The substance is also extremely reactive, and high concentrations can cause irritation of the eyes and mucous membranes – and, at worst, severe respiratory problems.

14

Interview with a Billy-Goat: The Smell of Denial

When you think about it, the human nose is a pretty odd-looking organ. Like an apparently random growth in the middle of our face, it looks almost as unnatural as a tail. In the case of Sigmund Freud, the father of psychoanalysis, it was an uncommonly voluminous body part and, since noses grow with age, it had become both fleshy and shapeless by the end, something like a root vegetable or a muzzle. What's more, it was always giving him trouble – breathing difficulties, unwanted dripping and a tendency to chronic blockage caused by swollen membranes.[1]

A friend of Freud's, the ear, nose and throat specialist Dr Wilhelm Fliess, carried out various surgeries on him, including an operation on his turbinates – the vertical, mucus-coated bony ridges that regulate the heat and humidity of inhaled air. But it did no good – quite the opposite. Perhaps it didn't help that Fliess also prescribed cocaine as both medicine and anaesthetic. If we add to all this a lifetime of endless chain-smoking, it's a wonder Freud's sense of smell worked at all.

He must have hated his nose and there is little evidence to suggest that he saw any use for it whatsoever, other than as a place to perch his spectacles.

The Psychologists

When the miasma theory died out, much of the interest in the relationship between humans and nature waned, and the sense of smell was no longer seen as necessary to survival. For the scientific elites, the time had come to pick up the reins of the Enlightenment once more, and they all backed the British naturalist Charles Darwin – best known for his *On the Origin of the Species* – when he came close to suggesting that evolution had reduced the nose to the same obsolescence as the little toe.[2]

Freud soon followed up with a psychological explanation. The sense of smell was little more than an unnecessary vestige of the days when we went about on all fours.[3] Back then, the nose was much closer to the genitals, and the sense of smell, stimulated by the tang of menstrual blood and other genital secretions, was much more closely linked to reproduction. When humans began to walk upright, the distance increased. At the same time, the sense of sight had evolved further, making the sense of smell superfluous. For modern individuals, Freud claimed, an excessive interest in smell could simply descend into neurosis.[4] If, in addition, the sense of smell was particularly sharp, it could end in hysteria – a condition primarily found in women.

The British psychiatrist and sexologist Henry Havelock Ellis thought along exactly the same lines as Freud:[5] the sense of smell was symptomatic of the twisted and perverted.[6] However, he added consolingly, the average Westerner need not despair if he occasionally succumbed to the influence of smell. It took more than that to be declared a deviant.[7] The German psychiatrist Richard von Krafft-Ebing, who, incidentally, regarded homosexuality as a pathology, then came up with a diagnosis he called olfactophilia – an

PETROL AND EXHAUST FUMES

ODORANT	GROSS FORMULA	MOLECULAR MASS [U]	BOILING POINT [°C]	ODOUR DETECTION THRESHOLD [PPB]
Benzene	C_6H_6	78	80	1,500–4,700
Toluene	C_7H_8	92	111	500–5,000
Nitrogen dioxide	NO_2	46	21	20–500

A petrol station is a whole world of odours, and among the hydrocarbons that make up the primary components of petrol and diesel, benzene was long the dominant odorant. With its simple, sharp, sweet aroma, almost like sour hard candy, it creates childhood memories for most of us, usually the good kind. But benzene attacks the nervous system too, as well as being carcinogenic and harmful to the reproductive system. It was eventually replaced by a less hazardous alternative, toluene, which has a somewhat higher boiling point but still has a sharp, sweetish smell. Like benzene, toluene can damage the nervous and reproductive systems, although the doses must be larger. Its vapour can also cause allergic reactions.

excessive interest in the smells of sex – which is still in the textbooks to this day.

Civilization

Utilitarianism was a branch of moral philosophy that increasingly gained favour over the course of the nineteenth century. Its central message was the greatest possible happiness for the greatest possible number of people. To this end, the individual had to subordinate themselves to the collective in every area of society, and individuals were useful only insofar as they were adaptable. This meant that

In the case of benzene, the threshold limit value beyond which it is harmful to health is lower than the odour detection threshold, which makes it especially scary. People will usually pick up the smell of toluene well before it reaches concentrations that are harmful to the health.

When fuel burns, it emits exhaust fumes. A sweetish hint may linger if combustion is incomplete, especially when the car is cold. Beyond that, the smell will be sharp, piercing and metallic, with a touch of chlorine and perhaps a whiff of burnt baked goods. That stems mostly from the nitrogen oxide emissions, which start off odourless and colourless but quickly react with the oxygen in the air to produce bluish nitrogen dioxide. Diesel exhaust contains more nitrogen oxides, so the smell is sharper and a touch more 'poisonous' than the odour produced by petrol engines. If you also detect the rotten egg smell of sulphur compounds, that means you have a defective catalytic converter.

It seems hardly necessary to dwell on the health implications of exposure to nitrogen dioxide, which are well known to most people – especially those with allergies or asthma. The threshold limit value is roughly the same as the odour detection threshold, so there is good reason to steer clear of even the slightest whiff of exhaust fumes.

individual sensory experiences had to be tamed and, to some extent, suppressed to ensure the steady march of human progress. Any tendency to glorify smell was taken to indicate regression to earlier and more primitive stages.[8]

This didn't go down well with the Decadents, of course, who claimed that happiness was a debatable and highly personal affair. But by the turn of the century their sense-driven strategy had lost much of its appeal, even among the bourgeoisie. This was further reinforced by the advent of the materialistic regimes of capitalism, social democracy and communism, which now sprung up east and west. It had become downright irrelevant to ascribe any decisive significance to the unconditional, individual sensory experience in the face of the real world.[9] The whole business was banished to the realm of art.

Oiled Tramlines

In the early years of the twentieth century, a wave of optimism and faith in the future surged through Europe. Vast amounts of concrete were ploughed into the further expansion of the metropolises, and people moved in from the countryside in droves.

In the villages they abandoned, most people had been on nodding terms with each other, or would stop for a chat whenever they met on the street and they might even have touched – their handshake accompanied by that tiny waft of scent. In the big cities, everything was alien. Contact was contingent on the sense of sight which was, in turn, flooded with stimuli along the thronging pavements and on public transport, face-to-face with others for minutes or hours at a time.[10] What little human odour remained was the lowest common denominator of the collective smell.

At the same time, the newcomers encountered a whole new world of aromas. Networks of electric trams sparked as they ran along oiled tracks, and the number of vehicles with internal combustion engines was constantly on the rise. Streets that had once steamed with horse dung were now enveloped in bluish-grey exhaust fumes, which would soon become the background smell in every European city. It was sour and acrid, it stung the eyes, but it was unquestioningly accepted as a necessary condition for the functionality of city life.[11]

Light, Air and Hygiene

The first suburbs, which soon appeared on the outskirts of the metropolises, were built according to the principle that form follows function – perfectly in step with the prevailing materialist ideals. In other words, simple geometrical structures, straight lines and clean surfaces, all executed in chalk white. The great ideologue of functionalism, Switzerland's Le Corbusier, provided the intellectual foundations: 'Architecture is the masterly, correct and magnificent play of masses brought together in light,' and 'I am and I remain an impenitent visual – everything is in the visual.'[12] Consequently, the construction of the new cities, like modern social intercourse, prioritized the sense of sight. Qualities related to sound and touch – resonance, timbre, texture and temperature – were toned down, while smell-related aspects were banished altogether.

Futurism

Back in the 1880s the uncompromising German philosopher Friedrich Nietzsche had hailed the sense of smell as 'our most refined instrument' – more sensitive than the spectroscope.[13] He inspired a whole new generation of artists who called themselves Futurists. They celebrated modernity and were unreserved technological optimists. Their Italian founder, Filippo Tommaso Marinetti, claimed to have found inspiration during a car crash amid the aroma of oil and puddles of petrol.[14]

The Futurists, who established themselves first in Italy and Russia, then later in England, claimed that intuitive knowledge acquired through sensory experiences was the proper basis for developing a deeper and truer understanding of the world – a perspective they adopted more or less wholesale from the Decadents.[15] But there the similarities ended. While the Decadents were steeped in pessimism and nostalgia, the Futurists claimed that a thorough study of the sensory depths was crucial if humanity was to break free of the past and enter the modern age. And this rupture must encompass society as a collective whole, whether people wanted it or not. The message of the Futurist Manifesto of 1909 was clear: 'The scent, nothing but the scent! That's all an animal needs.'[16]

The futurist Ennio Valentinelli elaborated on this in the manifesto entitled *L'arte degli odori* (The Art of Smells) from 1916:

> We must refine our nostrils. We have to start conquering the senses that have been elusive till now, we have to assimilate a new world and create a new lyricism. We have to force the reluctance of the senses when it comes to stench, we have to imprison, withhold, and enjoy, overcome nausea.[17]

A Feast of Scents

In practice, much of this resembled the caprices of both Huysmans and Montesquiou-Fézensac, but the experiments were on a far more magnificent scale. Some of them are described in a special Futurist Cookbook whose motto was 'No more pasta', since it only led to 'pessimism, nostalgic inactivity and neutralism'.

In a recipe for an 'Extremist Banquet', no food whatsoever is eaten. The event takes place one August evening, with five women, five men and one 'neuter', 'on a tongue of land dividing the most lake-like of lakes – slow, lazy, deserted, stagnant – from the widest and most marine of seas'. By way of an entrée, the guests recite a poem entitled 'Interview with a Billy-Goat' by Mario Carli. Then steaming food sculptures are trundled through the room, including a ship made of roast aubergines and strewn with vanilla, acacia flowers and red pepper. After three kitchen boys dressed in white silk and tall chef's hats have stormed in and proclaimed 'You are the bosses but you're rascals too . . . Stop all this mumbling or we'll boot you,' the French windows are opened onto the landscape outside:

> the first on to the mass of smells from the lake, the second on to the mass of smells from a barn and its associated storehouse of fruit, the third on to the mass of odours from the sea and its associated fish market and the fourth on to a hothouse and its carousel of rare, odoriferous plants.

The smells, 'curvilinear and succulent', tumble into the dining room and 'gratify all eleven starving palates'.[18]

Nougat and Gun Cotton

The Futurists' experiments usually involve aspects of catharsis and brutality. In the recipe for an Official Dinner, where the guests start by telling each other obscene jokes, the main dish is 'a multi-coloured castle of nougat with, inside, very tiny nitro-glycerine bombs which explode now and then, perfuming the room with the typical smell of battle'.[19]

The Futurists' sensual boldness also embraced the violent sides of existence, especially war:

> War is beautiful because it unites gunfire, cannonades, ceasefires, and the perfumes and odors of corruption into a symphony . . . Poets and artists of futurism . . . remember these

> principles of an aesthetic of war so that your struggle for a new poetics and a new sculpture . . . may be enlightened from within.[20]

Although at first glance the Futurists appear to be prejudice-free idealists on a ruthless quest for truth, we spy a particular obsession with the technological, the aggressive and the explosive. War and machinery spark far greater enthusiasm than the aromas of forests and human skin. The fact that future fascist dictator Benito Mussolini also disliked pasta was probably *not* the main reason why he was also keen on the Futurists.

15

The Battlefield: The Smell of Insanity

> We had come from lecture halls, school desks and factory workbenches, and over the brief weeks of training, we had bonded together into one large and enthusiastic group . . . We had set out in a rain of flowers, in a drunken atmosphere of blood and roses. Surely the war had to supply us with what we wanted: the great, the overwhelming, the hallowed experience. We thought of it as manly, as action, a merry duelling party on flowered, blood-bedewed meadows. 'No finer death in the world than . . .' Anything to participate, not to have to stay at home![1]

In his memoir, *Storm of Steel*, Ernst Jünger introduces us to the young men who marched into the First World War – himself among them. As inexperienced fledgling soldiers, they found themselves in a state of constant euphoria almost worthy of the Futurists. What awaited them out there was not just insane brutality, though, but a whole new world of odours too.

Ammunition

Wars have always smelt. Along with the reek of unwashed soldiers in damp uniforms and worn-out boots of leather and rubber, the sharp smoky smell of burning fields and towns has accompanied most military campaigns since antiquity. During the Renaissance, these smells were joined by the sulphurous gunpowder in rifles, muskets and cannons, with its sour fug of rotten eggs. Later, in the twentieth century, the dominant smell became the sharp, dusty stench of cordite from the new types of ammunition based on nitroglycerine and cellulose nitrate. And during the wars in Malaysia, Vietnam, Korea and Iraq in the latter half of that century, countless napalm bombs were dropped, with their dense stink of petrol and washing detergent. 'I love the smell of napalm in the morning,' declares Bill Kilgore as he lights a cigar in *Apocalypse Now*. It smelt of victory.[2]

At the same time, soldiers had been packed into constantly updated military technology – tanks and bombers and submarines. They smelt of musty locker-rooms mingled with motor oil, chemicals, cigarette smoke, rancid socks and collective sweat.[3] And the smell lingered when the crew went home on leave. 'I can recall returning to port on more than one occasion and watching my wife pinch her nose in disgust,' one submariner reported.[4]

Battlefield

Some way into the battle comes the raw, iron-tinged stink of blood from the wounds of the dead and injured. That goads the combatants into summoning up a strength they never knew they possessed. When calm returns and the blood has dried on clothes and soil, the insects swarm and decay soon sets in. The corpses change colour, first from white to greyish or greenish-yellow; soon after, to black, almost like tar. At the same time, they expand, stretching their uniforms, until all that remains 'is a vast, stinking, anonymous mass of blackening bundles'.[5]

During the U.S. Civil War, nurse Cornelia Hancock reported from the Battle of Gettysburg in 1863: 'A sickening, overpowering, awful stench . . . grew heavier and fouler until it seemed to possess a palpable horrible density that could be seen and felt and cut with a knife.'[6] But the soldiers had grown used to it. Corporal William Westervelt wrote letters to his family while eating breakfast alongside rotting piles of corpses: 'A few months before I could not have taken a mouthful of food and swallowed it in the presence of the corpse. Now, although they showed unmistakable signs of decomposition, we did not mind it.'[7]

Total War

As the wars raged, civilians sought refuge in air-raid shelters where they listened out for bombs – muffled bangs, like someone bursting gigantic paper bags. Down there, it smelt of damp bed linen, food hampers, pets, and women, children and old people who had slept in the same clothes night after night. And if the drainage system broke down, everything soon stank of excrement and old urine. It could get so bad that soldiers on leave insisted they preferred the trenches.[8]

And when the air raids were over, people climbed back up into the streets to be met by smells of plaster dust and sewage. According to survivors of the atom bombing of Nagasaki, it also smelt of burnt rubber and a nauseating electrical odour, like dried octopus.[9]

Two, three or four days later – depending on the temperature – the stench of dismembered corpses would emerge from the ruins. For the American writer Kurt Vonnegut, who was a prisoner-of-war in Dresden when the Allies bombed the city towards the end of the Second World War, it was reminiscent of roses and mustard.[10]

CHLORINE

ODORANT	GROSS FORMULA	MOLECULAR MASS [U]	BOILING POINT [°C]	ODOUR DETECTION THRESHOLD [PPB]
Chlorine gas	Cl_2	71	−34	200–400
Trichloramine	NCl_3	120	71	2–20

Chlorine is an element that does not exist in its pure form in nature but is produced industrially via the electrolysis of ordinary table salt, or the slightly bitter salt, potassium chloride. Heavy, yellowish-green chlorine gas, with its unique smell, can form even in sub-freezing temperatures and has an odour that we perceive as intense, sharp and clear but also slightly flat and simple.

In swimming pools disinfected with chlorine – as most are – you'll have smelt roughly the same odour of chlorine gas as the foot soldiers did when they came under a gas attack in the First World War. But it isn't exactly the same. Because in swimming pools much of the chlorine reacts with nitrogen compounds in the water, including sweat and urine, to form chloramines – primarily trichloramine. This substance has a lower odour detection threshold and smells less intense than pure chlorine, though it is also heavier and sweeter, often with a hint of urine.

Even in moderate doses, trichloramine can irritate the eyes and upper respiratory system, triggering asthma in those who are allergic to it. Nonetheless the health impacts of pure chlorine gas will be far worse. Yet despite its intense smell, very large doses are generally required for the concentrations to become genuinely dangerous. When that happens, the gas first attacks the trigeminal nerve, triggering a reflex that makes you want to hold your breath. In the end, you can't help gasping and the gas is drawn deep into your throat and lungs. Here it reacts with the body's moisture to form hydrochloric acid, which attacks the tissue and gradually blocks the oxygen flow. Every second it becomes harder to breathe. You quite literally drown on dry land.

The Trenches

When Jünger and his comrades marched into the First World War, they quickly found themselves embroiled in more or less static trench warfare – constant shellfire, coloured flares, chattering machine gun fire and a muddy no-man's-land scattered with fallen comrades and barbed wire.

The trenches, which were filled with rainwater at regular intervals, stank of faeces and rat carcasses, but the young recruits quickly grew accustomed to it. Gradually they even became used to the odour of corpses: 'It belonged to there. Moreover, this heavy sweetish atmosphere was not merely disgusting; it also, in association with the piercing fogs of gunpowder, brought about an almost visionary excitement, that otherwise only the extreme nearness of death is able to produce.'[11] But gas attacks were a different matter altogether.

Mustard and Horseradish

Poison gas was first used in the First World War. It started with chlorine gas in spring 1915. With favourable winds, it would sweep over the enemy lines like a yellowish-green cloud, 15 metres (50 ft) and 3 kilometres (2 mi.) long. The trenches were soon filled with the piercing smell of bleach, although some said it was more like pepper and pineapples.

Chlorine gas is an asphyxiant that works by paralysing the respiratory system. It was soon followed by nerve gases, blood agents and blister agents, such as mustard gas, which smelt of horseradish and mustard, phosgene gas – rotting fruit and mouldy hay – and Clark II – onions and bitter almonds. And Lewisite, which wasn't used in the First World War admittedly, had an unmistakable scent of geranium.[12]

It was almost like going into a well-stocked flower shop, or a shabby greengrocer if you prefer. These weren't smells that immediately prompted thoughts of danger. For the average soldier, the smells were more likely to suggest that the field kitchen was on its way. He simply had to learn the odours to understand what lay in wait.

Cyanide

During the Second World War, poison gas wasn't used in the warfare itself. However, massive amounts were used in the German concentration camps, primarily Zyklon B, whose active agent was hydrogen cyanide. Former SS guard Oskar Groening, who was tried for war crimes at the age of 93, describes the effect: 'The brain and the heart are first attacked. It begins with a stinging feeling in the chest, then it can cause spasmodic pain — similar to epileptic seizures. Death by cardiac-circulatory arrest occurs usually within seconds.'[13] The smell of cyanide reminds most people of marzipan and bitter almonds, but for genetic reasons, up to 30 per cent of all men and 10 per cent of all women cannot detect it at all.[14]

The Crematoria

For the prisoners destined for the gas chambers, the odour from the crematorium ovens was far more present:

> The smoke did not always rise above the chimney in transparent, bluish clouds. It was sometimes pressed down to the

> ground by the wind. And then one could notice the unmistakable, penetrating stench of burnt hair and burnt flesh, a stench that spread over many kilometres.[15]

The intense odour that came in sporadic waves at all times of day and night also affected the local farmers. They guessed what was going on and rarely accepted the lie that it came from the camps' bakeries. And of course, the German personnel cannot have failed to notice it: the smell was ubiquitous. But since there was little to be done about it, it soon became a non-subject – a taboo.[16]

In his journal, Johann Paul Kremer, one of the doctors at Auschwitz, gives a detailed account of his meals: 'Excellent food in the Home . . . we had sour duck livers for 0.40 mark, with stuffed tomatoes, tomato salad, etc. Water is infected so we drink seltzer water which is served for free.'[17] On 25 October 1942 plum brandy was on offer, and on 2 November he drank Bulgarian wine and was allowed to eat as much as he wanted. All this is described in painstaking detail. But neither he nor any of the others once mention the smell, although it must have been overwhelming.

The Blonde Angel

The Nazi regime was obsessive about cleanliness and hygiene. Dirt and body odour were associated with physical and moral decadence. In Hitler's manifesto, *Mein Kampf*, his attitude to the Jews is absolutely unequivocal:

> The cleanliness of these people, moral and otherwise, I must say, is a point in itself. By their very exterior you could tell that these were no lovers of water, and, to your distress, you often

> knew it with your eyes closed. Later I often grew sick to my stomach from the smell of these caftan-wearers.[18]

In the overcrowded cellblocks, the terrible lack of hygiene and brimming latrines created an overwhelming background stench that matched even Hitler's most grotesque fantasies. But the prisoners had largely become accustomed to these conditions. Only an unexpected waft of perfume from a prison guard could rouse their sense of smell anew.

In Auschwitz, the twenty-year-old SS guard Irma Grese regularly paraded between the rows of bunk beds to select new victims for the gas chambers. Known as 'the blonde angel of death', she was so beautiful and sweet-smelling that even the prisoners were captivated. One survivor later described her as an almost divine revelation:

> Wherever she went she brought the scent of rare perfume. Her hair was sprayed with a complete range of tantalizing odors: sometimes she blended her own concoctions. Her immoderate use of perfume was perhaps the supreme refinement of her cruelty. The internees, who had fallen to a state of physical degradation, inhaled these fragrances joyfully. By contrast, when she left us and the stale, sickening odor of human flesh, which covered the camp like a blanket, crept over us again, the atmosphere became even more unbearable.[19]

During the trials when the war had ended, several of her victims spoke of the smells that had impregnated the concentration camps. They said that similar smells could still revive those appalling memories, like a ghastly variant of Proust's madeleines.

Reactions like these have been reported from other wars too. After the bloody civil war in Cambodia in the 1970s, many people found that even the most ordinary smells, like exhaust fumes, cigarette smoke, rubbish, nail varnish, fresh paint and perfume, could provoke paralysing panic attacks.[20]

16

Love Apples: The Smell of Lust

The Wall Street Crash of autumn 1929 came after several years of surging economic growth and marked the start of the Great Depression. On the morning of Black Thursday, stockbrokers had arrived at the temple-like Stock Exchange as usual. Neatly combed and bursting with arrogance, they were ready for action. And, as usual, the scent of fresh forest must have wafted through the hallways. Not because of the decorative foliage in the foyer but because all the brokers were pumped full of the male hormone, testosterone.

Testosterone

Testosterone contributes to reproductive capacity, beard growth and deep voices. Some men secrete more than others, either because of their age – levels peak in the twenties – or simply because they happen to produce more of the substance. The average stockbroker is probably at the high end of the scale, because there's a clear link between testosterone levels and risk taking.[1] And sang-froid that borders on recklessness is almost a prerequisite for success in the sector.

As at the Stock Exchange, a haze of testosterone will always be present wherever the stakes are high: in boardrooms, government offices and, above all, wartime command centres where the old

SEMEN

ODORANT	GROSS FORMULA	MOLECULAR MASS [U]	BOILING POINT [°C]	ODOUR DETECTION THRESHOLD [PPB]
1-pyrroline	C_4H_7N	69	88	low[2]
Androstenone	$C_{19}H_{28}O$	272	371	0.2

Fresh semen, a slightly viscous fluid, is made up of sperm cells that are dispersed in the seminal fluid. It emits a muted and somewhat flat smell of ammonia and leavened dough with a hint of chlorine. That is the odour of a substance called 1-pyrroline, which forms when the components of semen – the nitrogen compounds spermine, spermidine and putrescine – come into contact with air, oxidizing almost instantly.

The rest of the smell of semen stems from the steroid androstenone, a male sex hormone that is closely related to testosterone. Even though it has an extremely low odour detection threshold, only half of all men and 90 per cent of women pick it up. And those who *do* smell it report startlingly divergent associations.[3] Some claim it smells of warm musk and acrid urine, while others will describe it as the sweet aroma of freshly cut grass, flowers and vanilla.

Synthetic versions of androstenone are produced for use in animal husbandry. Like most mammals, the pig's nose is equipped with a vomeronasal organ, which is directly adapted to translate androstenone into arousal. Although humans also have this organ, it is severely underdeveloped, and may not have any function at all.

generals are planning their next move. But in this last case, the odour will have lost much of its edge, supplanted by a smell more reminiscent of old cooking oil and rancid grease – or at best, with a bit of goodwill, a little like moss. Because the freshness wanes with age.

The Crash

Stockbrokers are usually young and virile, and on the morning of Black Thursday it must have smelt like a hike in lush countryside. Until everything suddenly went pear-shaped and the stock market dropped like a stone. Panic ensued. The bold aroma of testosterone was replaced in an instant by the reek of stress sweat pumped out by the apocrine sweat glands in the armpits and groins and around the anuses of the distraught stockbrokers. Fresh sylvan nature was suddenly transformed into a busy industrial kitchen full of steaming saucepans of meatballs and onion.

It's a hard smell to cope with. Research shows that even experienced dentists may end up delivering much poorer treatment after exposure to the smell of frightened patients.[4] And the fear must have spread through the Stock Exchange like wildfire. Only the toughest brokers, and perhaps one or two who happened to have blocked noses, will have managed to keep their heads.

Despite the disdain of Freud and the other medical experts of the day, then, we can assume that the sense of smell continued to wield its influence into the first half of the twentieth century. And the effects weren't limited to the corridors of power but affected all social arenas down to the intimate sphere and sexuality.

Vagina

In the hardscrabble 1930s that followed the Crash, a breath of broad-mindedness and erotic passion wafted through the Western world. Henry Miller's *Tropic of Cancer* fairly steamed with sex: 'sap still oozing from between her legs, a warm feline odor and her hair in my mouth. My eyes are closed. We breathe warmly into each other's mouth.'[5] A woman's vagina contains a multitude of smells.[6] In *The Vagina Monologues*, author Eve Ensler lists the following: earth, wet garbage, sweet ginger, sweat, pineapple, earthy meat and musk, deep, deep forest, damp moss, peaches and somewhere between fish and lilies.[7]

She doesn't mention a 'warm feline odour', and besides, Miller is almost certainly describing a vagina that has recently played host to a man's semen – a substance that many would claim has a dense, animal scent of its own, of musk and urine. Even without a dash of woman, it can be intense. 'I fear from what I see and hear that perverse sexual practices are again getting the mastery over him,' said prison chaplain W. D. Morrison of the Irish author Oscar Wilde who was convicted of homosexuality in the 1890s. 'The odour of his cell is now so bad that the officer in charge of him has to use carbolic acid on it every day.'[8]

Genetic Guidance

The sense of smell is not a panorama sense like sight. The olfactory identity each of us carries around with us is mostly shared with those who are physically close to us, and provides information about, say, arrogance and stress. It can also communicate arousal, without necessarily making other people feel aroused.

One important group of these signal substances, which also include androstenone and testosterone, is pheromones. They are said to prevent interbreeding, because we humans have a preference for the smell of partners whose immune systems differ from our own.[9] So when we raise our arms and unleash a stream of aromas, we are not just ensuring that the other person's potential as a partner becomes clear more quickly but securing the health of our descendants. Unless, that is, we are stuffed full of hormone products that can quickly muddy the waters.

So far, all this takes place at the unconscious level, although the knowledge may well spawn a more or less clear gut feeling. Next, the mapping of the partner is supplemented with more tangible information. Gonorrhoea smells fungal, even in the armpits, and if a vagina or penis smells of fish, that may mean a case of chlamydia.[10] For the initiated, lung cancer, gum disease, multiple sclerosis and certain mental disorders such as schizophrenia can reveal themselves via particular odours.

Disease

The idea that diseases affect our biochemical processes and alter our body odour was common knowledge as far back as the times of the ancient Greeks. And Hippocrates, the father of medicine, viewed the nose as the most useful of all diagnostic tools: physicians would systematically smell their way through the patient's breath, skin, flatulence, faeces and urine, sometimes even conducting a taste test.[11]

Even minor nuances were crucial, but roughly speaking, plague smelt like old apples, typhoid fever like freshly baked bread, tuberculosis like sour beer, yellow fever like the butcher's shop, urinary tract infections like caramel, smallpox like freshly peeled onions, kidney disease like old hay, and leprosy like a mixture of goose feathers and goat. And the breath of diabetics smelt of lemon or rotten apples, tipping over into acetone in the later stages.[12]

This diagnostic practice persisted into the late eighteenth century, when miasma theory put a stop to it all.[13] The boundless fear inspired by the stink of decay and decomposition even led to evil-smelling patients being turned away from the hospitals and left to fend for themselves.[14] Once the hysteria died down towards the end of the following century, the method enjoyed a certain upswing, especially among physicians in remote districts, who had little else to rely on.[15] But as the sense of smell was downgraded in society as a whole, it was soon supplanted by visual assessments, along with the use of tests and instruments.

Scented Gifts

In all the history of humanity, our intimate scent signals can rarely have been assigned less weight than they are today – not just in medicine but the erotic world too. In the sixteenth century, people exchanged love apples. A woman would carry a peeled apple in her armpit until it was saturated with sweat, then give it to her beloved as a scented gift.[16] And a hundred years later Napoleon wrote to his wife Josephine, urging her not to bathe until he returned home. He was planning to sniff his way through the entire arsenal of her natural aromas.[17]

Today that is mostly seen as a slightly icky anachronism. Despite Henry Miller's lush literary campaigns, the status of natural body odours has sunk to ever lower levels in modern eroticism. Indeed, every effort is made to eliminate them through conscientious cleaning, intensive scrubbing and endless showering.

Whatever the health-related arguments in favour of this practice, they line up poorly with the facts. Just ask any polar explorer who's gone for months without any contact with water. No matter how refreshing it may be, we don't actually *need* to bathe. The human body is, in principle, self-cleansing. Most dirt and grime is naturally excreted via fatty substances in the skin and continues its journey out into the world via the thousands of skin cells we shed every day.

Pubic Hair

And yet nowadays we bathe more than ever before. What's more, more than half of all European women and one-third of men regularly remove their pubic hair.[18] Barely a century old, the practice unsurprisingly coincides with Freud's general rejection of the sense of smell, and Darwin's theory that sparser body hair indicated a higher level of development.[19]

Pubic hair traps and retains the sexual aromas almost like a data warehouse and, given the ample access to heat and moisture, the scents are further cultivated into lush and complex blends. The aromatic result was highly prized well into the nineteenth century, when even snobbish British aristocrats would trim their hats with a twist of their mistresses' pubic hair.[20] George IV of England – Napoleon's contemporary – left behind a snuffbox stuffed with pubic hair, which he undoubtedly took out in solitary moments.

Yet Norwegian sociologist Lise Widding Isaksen isn't quite right to declare that 'today's aesthetically and politically correct body is smooth, muscular, lubricated and odour-free.'[21] For even in intimate relations, it is acceptable to use camouflaging scents. This practice was previously preserved for public spaces, where the bourgeoisie had for centuries used perfumes to avoid being mistaken for emissaries of the Devil, members of the working classes, menstruating women or whores. Even Napoleon doused himself in two bottles of eau de cologne per day to secure the respect of his underlings.[22]

Deodorants

But with the advent of the twentieth century, public and private olfactory spheres soon merged. And the role previously played by perfumes in exclusive flacons was fulfilled by simple mouth sprays, deodorants, and scented soaps and shampoos for daily use. A 1920 advert for the German mouthwash Odol makes it clear that women were the primary target:

> We adore the blonde gentleness of the German woman, the cool elegance of the Scotswoman, the grace of the Frenchwoman, the bright freshness of Scandinavian femininity and the springy resilience of the woman of the United States . . . But regardless of all differences of racial type, we adore a woman with a fresh mouth, beautiful, snow-white teeth and seductively pure breath.[23]

Eventually, men were also fed the idea that a clean, fresh body odour was almost the price of admission to modern life – almost a prerequisite for having any human contact whatsoever. There's 'always room for one more as long as they use Rexona', declared a cinema advert for deodorant over a clip of an overcrowded tram; and many of us who grew up in the 1960s and '70s remember how generously our fathers drenched themselves in Old Spice aftershave.[24] The extraordinarily dense and cloying scent, which came in a white bottle with a yacht motif, cast its pall over the toast and marmalade at many a breakfast table. And it was supplemented by mouthwash, which generally had a strong, fresh aroma of menthol and eucalyptus.

Sex Pheromones

In the 1990s the perfume segment encroached even further on the heart of eroticism, no longer content to merely camouflage smells. Now, women were also offered a battery of chemical pheromones that would boost their sex appeal and make them irresistible to men. Vanilla, amber and rosemary scents were blended with copulins – whose name is derived from the word copulation – supposedly the counterpart of male androstenone. The products caught on and alternatives for men soon followed, with added synthetic androstenone – so-called Alpha-Male Concentrate – supplemented by heavy doses of leather and musk.[25]

All these are expensive products whose effects are very poorly documented.[26] And if sex pheromones *are* significant, we might as well let nature do the job: after all, assuming these signal substances exist, they must be plentifully present in our own body chemistry. But if we do decide to leave it up to nature, we're also going to have to accept that we need to give pheromones a freer rein instead of eliminating them with frenetic intimate hygiene.

17

White Sheets: The Smell of Purity

In the book *Anarchy in the UKR*, Ukrainian writer Serhij Zjadan recalls the post office of his childhood with its 'quite remarkable odour, I can smell it to this day . . . the smell of silicon glue, the smell of fresh newspapers, the smell of sealing wax, lumps of dark-red lacquer that shone on the letters like tiny suns'.[1] And if we add to that the sweetish-sour notes of the ink-pad, fresh brown wrapping paper and the reddish-orange, water-sodden rubber sponges by the counters, we find ourselves in an olfactory world that will prompt a nod of recognition from anyone who grew up in the latter half of the twentieth century.

Cold War

The Cold War had divided Europe, and east of the Iron Curtain one particular smell really stood out: the smell of chlorine. The aromas of black bread, sheepskin, vodka and unwashed bodies that once typified pre-Revolutionary Russia had long ago been eliminated.[2] Now the smell of chlorine was everywhere – in industrial workplaces, children's nurseries, railway stations, hospitals, dairies, butchers' shops, offices and public baths, not to mention canteens and public eateries – although there, it was usually blended into a distinctive odour that also incorporated boiled cabbage.[3] In the USSR, everything was washed and disinfected in a communal effort that bordered on the ideological. A Communist must always be in good shape, of course – but above all, clean.

West of the Iron Curtain, those lucky enough to have relatives in the USA received exotic Christmas parcels from America. Packed with assorted hard candy, nylon stockings, coffee, clothes, raisins, canned goods, dolls, chocolate and chewing gum, they exuded a heavy aroma of sweets and mothballs.[4] It was the smell of capitalism.

In the 1950s more and more western Europeans are accommodated in well-organized white suburbs. They are brought up according to functionalist ideals, and the bright airy flats they live in are divided into well-defined spaces. First the vestibule – the sluice between the outer and inner world – where dirt borne inside on shoes and outerwear is kept at bay. Then onward through the living room and into the kitchen, where the meals are prepared and eaten. Potato peel, egg shells and all the other food refuse is dispatched through the rubbish chute and any smells that might arise are whisked away by the kitchen fan before they have a chance to spread. But these are mere trifles anyway: ready meals and canned food have now made their appearance and, other than a spot of light baking, very little substantial preparation is done. Jointing meat on the kitchen counter, pickling, drying and salting are all chapters that closed long ago.

And now even the smallest flats have their own bathroom and WC. All filth and fluids are continuously flushed out of our buildings and away through underground drainage pipes to who knows where – out of sight, out of mind.

Dust on the Brain

As doorbells chime, record players spin and radios play – all day long, from *Thought for the Day* to the *Ten O'Clock News* – the hygiene drive forges onwards, full steam ahead, with much the same conviction and determination as east of the Iron Curtain. In Eva Ramm's 1958 book *Med støv på hjernen* (Dust on the Brain), a member of the self-appointed hygiene police in a group of housewives makes it clear what is expected:

> 'You can say what you please,' she said, 'but there's no getting away from one thing: A house has to smell clean, and it most certainly won't if you don't spend every second of spare time you have between childcare and cooking on washing and scrubbing.'[5]

The housewife was expected to be a straight-backed soldier in the war against dirt and disease, and any attempts at desertion were firmly suppressed. Her standing orders were to eradicate all dangerous bacteria, but since they were too small to be seen with the naked eye, the house could, quite simply, never be clean enough.[6] At least it could smell clean, though. For the idea of the disease-inducing stench was still alive and kicking in the last half of the twentieth century, almost a hundred years after the miasma theory was toppled.[7] For 1950s mothers, the dry, slightly scorched smell of freshly ironed and starched laundry straight from the linen closet was almost a weapon in its own right – like brandishing a crucifix or a rosary at a vampire.

As in the USSR, detergents were often based on chlorine, which dissolves grease. The smell was brutal, and lemon and lavender

TOBACCO

ODORANT	GROSS FORMULA	MOLECULAR MASS [U]	BOILING POINT [°C]	ODOUR DETECTION THRESHOLD [PPB]
Nicotine	$C_{10}H_{14}N_2$	162	247	50–2,000
2,3,5-trimethylpyridine	$C_8H_{11}N$	121	184	0.1–0.5
2-ethyl-3,5-dimethyl-pyridine	$C_9H_{13}N$	135	198	0.03–0.1
2,5-dimethylpyrazine	$C_6H_8N_2$	108	155	1–10
Acetaldehyde	C_2H_4O	44	20	60–8,000
Formaldehyde	CH_2O	30	−19	20–190
Acetone	C_3H_6O	58	56	20,000–65,000

When dried fermented tobacco leaves are lit, more than four hundred different odorants are released. The complex blend is often described as sour, dry, sharp, piercing and cloying, with hints of fish, hay and leather. On top of this come fumes from the smouldering paper, the cigarette filter and flavourings such as vanilla, menthol, sorbitol, rum and various perfumes.

While the nicotine, which contains nitrogen, contributes a mild, fishy undertone, the smell of tobacco is dominated by a few tarry pyridines, 2,3,5-trimethylpyridine and 2-ethyl-3,5-dimethylpyridine – which smell of both mould and fish – and a few pyrazines, primarily 2,5-dimethylpyrazine, with its sour, earthy notes. Although these substances are usually present in low concentrations, they largely compensate for that with their low odour detection thresholds. Even so, they are often drowned out by high concentrations of the sharp, sweetish oxygen compounds, acetaldehyde, formaldehyde and acetone, produced by both tobacco and additives.

The smell of smokers' breath that lingers in the oral cavity and the respiratory system is also dominated by pyridines and pyrazines. The same applies to the smell of old smoke that has settled in walls, furniture, carpets and clothing. In these cases, the volatile components have evaporated, leaving behind the heavier molecules that tend to form a sticky film over surfaces. The smell can often be eliminated through oxygenation with ozone gas.

odorants were soon added to the products to make the feeling of cleanliness more appealing. This meant that a heretical housewife could easily get away with spraying a spot of floral scent in the corners: it *smelt* good enough. And if she failed to achieve entirely satisfactory results in the bathroom, she could always resort to chemical odour-neutralizers with active ingredients that would break down any hint of sulphur that might be lingering after a hefty trip to the toilet.

Sports Halls and Smoking

Yet even in this sterile environment of white bed linen and homes that increasingly resembled hospitals, there were still some zones of olfactory freedom – at least if we are to believe Swedish author Mikael Niemi. In his autobiographical book *Popular Music from Vittula*, he describes the experience of a group of middle-aged women in a rural gym hall.

> Bum sweat cascaded over blubbery backs, the air was alive with a whiff of pussy. 'Up again, then left and right,' sideways thrusts with ample hips . . . The women fell like two-ton bombs, lay slithering in the pools of sweat on the varnished floorboards . . . The hall stank of marshy swamps and menopause. Death and birth in an ancient mix.[8]

With their grimy rubber mats and musty locker rooms, sports halls have always been a kind of safe space for smells and have proven impossible to fix. At home too, though, there was one smell that remained sacrosanct: the aroma of tobacco. Even the great ideologue of functionalism, Le Corbusier, was seldom seen without a straight pipe, and the suburban housewife in Eva Ramm's novel wasn't in the least bit worried about breathing in the clouds of smoke from her husband, who was puffing away on the sofa with his legs up on the cushions. Quite the opposite really: 'It smelt of man. And a woman should always have the smell of a man in her nostrils.'[9]

Tobacco served as a kind of secular incense, a pheromone-like expression of masculinity and power. The effect was usually intensified by seductive aromas of campfire, or a sailor's life – all ropes and tarred ships' hulls. And matinee idols like Marlon Brando and James Dean always appeared with a cigarette dangling out of the corner of their mouths.

Although the dark side of tobacco was already clear in the 1940s, it took several decades for that realization to become embedded in the culture. But once it happened, attitudes towards the smell of even the most aromatic cigar shifted altogether: it was now perceived as unpleasant. Today, few smells do more to reduce the value of a property than the stink of ingrained tobacco smoke.

Chipboard and Vinyl

In the 1950s the market was more than ready for new materials to replace wallpaper, plasterboard and wooden surfaces where dirt could still hide out in pores and cracks. And soon they came: laminates, wall foil, flooring and surface treatments that didn't just provide a smooth and complete-coverage finish, but were completely impervious to vapour.

The most popular products included high-pressure laminates (HPLs) produced by compressing patterned sheets of kraft paper and synthetic resin into panels. The brands included Formica, Arborite and Greenlam – just savour those names! Other innovations were chipboard, which was produced from woodchip soaked in formaldehyde resin, PVC-based vinyl covering for walls and floors, and paint with acrylic or artificial latex binders – all fruits of the young and rapidly expanding petrochemical plastics industry. And if any cracks appeared, they could easily be sealed with a large arsenal of spackling and jointing products – also based on plastic.

By the start of the 1970s the average Scandinavian apartment contained more than a ton of plastic. Before the turn of the millennium the level had rocketed to around 3 tonnes.[10] Nothing gets in or out through these materials. It's like the wildest dreams of your average nineteenth-century miasma fanatic come true at last.

But all this takes on an almost tragicomic edge when we consider that bacteria and viruses both thrive on plastic surfaces. While the feared COVID virus gives up the ghost after a mere 24 hours on regular wallpaper, it can survive for up to 72 hours on plastic wallcovering.[11]

New Car

Even perfectly conventional wooden houses no longer smell of timber. They do, however, smell a great deal of the new materials. Formaldehyde resin from laminate smells of gherkins with an added chemical twist; vinyl flooring of sweet petrol; and latex paint of sour milk, at least until it's fully hardened. But these smells are usually considered just as unproblematic as the new industrial odours were during the Industrial Revolution. They are sterile and chemical and cannot possibly be associated with bad housekeeping. First and foremost, they smell of progress.

This perception is so ingrained that many people are both disappointed and dubious when they find that their new car doesn't give off a hefty dose of the expected smell of, yes, new car – a combination of more than fifty different odorants from seat covers, inserts, insulating foam and fresh lubricant.[12] But there's a solution available to smart salesmen: Little Trees.

Little Trees car freshener, or Wunder-Baum, was developed by the German perfumier Julius Sämann. The prototype was inspired by a milkman who complained about the persistent stink of spilt milk in his van. Small pine-tree-shaped scraps of card dipped in synthetic perfume proved well able to bring the situation back under control. And in the Scandinavian '*råne*' subculture – centred on cruising around in pimped-up cars, and dominated by young men – the prevailing fug of hormonal testosterone and apocrine stress sweat was remedied by dangling entire bouquets of Little Trees from the rear-view mirrors.

The violet-coloured cardboard pine is the one that's impregnated with the scent of new car. And that Little Tree is also the car salesman's salvation. With the tree in place, he can expect his customer not just to be more trusting, but to be in a state of heightened euphoria and naivety when it's time to negotiate the price.

Immune System

These new materials made it far easier for conscientious suburban housewives to keep their apartments clean and tidy, especially with the aid of ever more effective detergents that cleaned 'like a white tornado'.[13] But even though we had now achieved 'the ideal hygienist dream', children who grew up in this period were more prone to asthma and allergies than any previous generations.[14]

In 1989 the British epidemiologist David Strachan proposed what he called the hygiene hypothesis: in order to develop a properly functioning immune system, the human body must be exposed to natural bacteria early on in life. Children who eat soil and grow up with cats and other pets are better protected; the more contact they have with the natural world and biodiversity, the better.[15]

The hypothesis, which is almost like a rerun of Renaissance notions about the healthiness of a dirty body, has since been confirmed by numerous studies: the bacteria cultures we are exposed to help create a kind of living armour between the human body and its surroundings, strengthening the immune system and making us more resistant to infections, inflammation and allergies.[16]

Nowadays, more than 40 per cent of the population of the Western world suffers from allergies, whereas the level rarely rises above 10 per cent in countries that escaped the hygiene hysteria of the Western world.[17]

18

The Blood of an Englishman: The Smell of the Unknown

'Fee, fi, fo, fum! I smell the blood of an Englishman!' roars the outraged giant in *Jack and the Beanstalk*. He knows what that means. And it doesn't look good.

Curry

Towards the end of the twentieth century, African and Asian immigration, formerly focused on the old colonial powers of France, the Netherlands, Belgium and Britain, began to spread further north in Europe. Many of the new arrivals were from Pakistan, Turkey and Morocco. And now, the few remaining odours that had somehow survived the West's prevailing obsession with smells – the faintest whiff of urine in a stairwell, say – were joined by myriad new and unfamiliar scents. Many of the original inhabitants were as outraged as the giant in the fairy tale. This smelt of the unknown and it smelt scary.

The aromas that filled the stairwells were mostly food-related: spicy curry blends, fermented fish sauce, cumin, fenugreek and garlic, which marked a dramatic departure from the wafts of potato, fricassee or fish balls that had occasionally escaped the Scandinavian kitchens. Once in a while there might also be a whiff of breath or

GARLIC

ODORANT	GROSS FORMULA	MOLECULAR MASS [U]	BOILING POINT [°C]	ODOUR DETECTION THRESHOLD [PPB]
Allicin	$C_6H_{10}OS_2$	162	248	0.2–1.5
Dimethyl sulphide	C_2H_6S	62	37	1–20
Allyl methyl sulphide	C_4H_8S	88	92	0.1–2.5

Garlic, *Allium sativum*, originated in southern Asia and exudes a slightly piercing, acidic, bitter and sulphurous odour when its cells are crushed by chewing, cutting or mashing. The smell stems primarily from a substance called allicin, although some sulphides are also involved, including dimethyl sulphide, which contributes a hint of boiled cabbage and seafood.

Once inside the digestive system, allicin is converted to allyl methyl sulphide. Unlike most other food molecules, this is a compound that the body cannot break down, so instead it is sent directly into the bloodstream and excreted via urine, breath and sweat. Here it combines with other odorants into a pungent, sweetish-sour and slightly metallic blend that can linger for hours, no matter how thoroughly we brush our teeth or bathe. After a few days of liberal garlic consumption – especially fresh garlic – the levels of sulphur compounds in the blood will build to new heights and the smell can become extremely intense, with hints of both sourdough and sharp cat pee.

Since the odour detection threshold is low, a little goes a long way, but we can diminish its intensity a bit by simultaneously eating foods containing enzymes that break down the sulphur compounds – like celery, mint leaves and basil. And it may help to wipe your armpits with vinegar.

sweat. And if the original residents unexpectedly ended up in a lift with one of the new arrivals, hints of the unfamiliar aromas might linger in their hair and clothes. The sum of all this was perceived almost as 'biochemical threats to the integrity of white bodies and minds'.[1]

Cow's Milk

Smell differences are not genetic. People all over the world smell pretty much alike in principle, although Japanese people have slightly fewer than average apocrine sweat glands, and Europeans slightly more.[2] But those are trivial differences. The main thing that sets us apart is food. Different cultures have markedly different diets, which means that the type and range of decomposition products that are transmitted via the blood to breath and sweat are also different.

And it goes both ways. An Asian or African will soon notice the distinctive smell of European and American sweat, caused by high consumption of cow's milk and butter. It has characteristic notes of vinegar and fish – so much so that during the Vietnam War, the Viet Cong could smell the Americans from a long way off, even in dense jungle. After the Americans became aware of this, newly arrived soldiers were immediately put on the local diet and their milk consumption reduced.[3]

The heavy element of sulphurous molecules in garlic quickly passes into the blood and is excreted via both breath and sweat. For Europeans, this had been a crucial marker of sinister, foreign odour since well before the garlicky immigrants came on the scene in modern times. As far back as the Roman Empire, the smell of garlic had been used to stigmatize the peasantry, while antisemites had, for millennia, linked it to the Jews – even claiming that the stench only abated after they had converted to Christianity.[4] Although garlic still hasn't gained full approval in northern Europe, it did, at least, lose that insinuating power after becoming a self-evident part of the Western diet.

The Smell of the World

When the first European globetrotters set out on their travels in the early twentieth century, they found that smells changed the further away they got from their homeland. British doctor Dan McKenzie was fascinated:

> Who can ever forget the bazaar smells of India, the mingled must and fust with its background of garlic and strange vices, or the still more mysterious atmospheres of China with their deep suggestion of musk? Naturally the air of a cold country is clearer of obnoxious vapours than that of tropical and sub-tropical climes, but in spite of that, the first whiff of a Tibetan monastery, like that of an Eskimo hut, grips the throat, they say, like the air over a brewing vat.[5]

The European adventurers also discovered that the sense of smell sometimes had a completely different status elsewhere than it did in Europe. In Arabic countries, people breathed on each other as a sign of friendship and goodwill.[6] And in India, sniffing another person's head was an expression of devotion: 'I will smell thee on the head, that is the greatest sign of tender love.'[7]

The Nose as Compass

Back home in Europe, anthropologists were busy ordering and categorizing the world's races by skin colour and skull shape. Some also studied the sense of smell, declaring that it was assigned greatest weight by indigenous people who lived close to nature – a clarification that shored up the preconception of the day: that smell was a raw and primitive sense.

Less prejudiced social anthropologists later offered more nuanced explanations. In Oceanic cultures where every single island has its own particular scent, the sense of smell is absolutely crucial for navigation on the open sea – almost like a compass.[8] And for tribes living in dense jungles, smell is more useful than sight for locating fires and prey and fruit up in the tree canopies.

The Ongee people on the lush Andaman Islands in the Bay of Bengal arrange the year into a calendar of smells based on the plants and trees that blossom at different times. When they meet, they greet each other by saying: *Konyune? Onorange-tanka* (How is your nose?).[9]

Homogenization

Baudelaire – the nineteenth-century French poet who was prosecuted for his steamily erotic poetry – was also fascinated by the smell of the wider world. Although he didn't go abroad himself, he wrote of others who had done so: about how Cape Town smelt of sheep, Russia of leather, the Tropics of coconut oil and the Orient of musk and corpses.[10] Half a century later, the British author D. H. Lawrence had experiences of his own with the smell of Mexico, which he described as a blend of resin, carnations, coffee, sun-burned earth and urine.[11] And I well recall my own experience in Athens, on my first Interrail trip in the early 1970s: the whiff of pungent cat pee, rotten tomatoes, chargrilled lamb and, of course, generous doses of garlicky sweat.

There's barely a soul who could boast of such experiences these days. Now, in the early part of the twenty-first century, Western culture has gradually tightened its antiseptic grip around the globe. All the airports, where the architecture, sounds and tastes are all the same, also smell alike. Even the smell of the centre of Calcutta, where the markets and bazaars are surrounded by elegant shopping malls, is now only marginally different from the smell of London, while the port of Cape Town smells almost indistinguishable from that of Copenhagen.[12]

But just a few stones' throws away from the impeccable avenues, the smells are still different, especially in the poor districts. In rural communities the difference is greater yet. For now. Because even in the Amazon Rainforest, change is afoot. Avon ladies are heading into the most remote villages in ever-increasing numbers, going door to door, bearing cartons packed with standardized fragrances. Sometimes the currency is local products: two dozen eggs for a Bart Simpson deodorant, and 10 kilograms (22 lb) of cassava flour for a small vial of a perfume called Charisma.[13]

Calvin Klein

Halfway through the nineteenth century, the French anthropologist Paul Broca announced that White people had bigger brains than Black people. The same went for men versus women and the upper classes versus the lower. He also claimed to have observed that Frenchmen had larger brains than Germans.[14] On that basis, he placed French culture at the very top of the hierarchy of civilization, cultivation and intelligence.

Smell has always been used in the same way throughout history. Aristocrats, clerics and both cultural and economic elites have placed themselves at the odour-free centre and, the further out to the periphery anything is, the worse it is deemed to smell. From this position, they have constantly sought to legitimize men's sovereign place in religion, the natural rights of the upper classes and the supremacy of Europeans.[15]

If you are constantly told that you smell bad, it isn't long before you start to believe it yourself. In European and American cities, immigrants and other minority groups soon set about adapting to the prevailing norms to avoid standing out. 'I want people to smell Calvin Klein and not my wife's curries,' admitted one man in an American survey.[16] And real estate agents in New York required Asian immigrants to take particular care ahead of viewings: 'There is nothing that will annoy a potential home buyer but to be met at the door with the smell of years of fried food and spicy cooking. The buyer will assume that the smell is permanent and cannot be scrubbed out.'[17]

Romany

The fact that some minorities refused to allow themselves to be gagged became a particular problem in the most thoroughly sanitized societies – like Sweden. Per capita consumption of cleaning products there stands at more than 10 litres per year, and people often shower twice a day.[18] In a survey, 33-year-old Fredrik firmly declares: 'If you've had a shower, you're safe . . . You know you're clean, you know you don't smell.'[19]

When a group of Romany travellers parked their caravans on a vacant lot in Malmö in 2014, the neighbours were quick to protest. What bothered them most was the smell, which they described as sharp, nasty and nauseating: 'ordinary people in their homes must have access to fresh air when they, for natural reasons, open their windows.'[20] After a few days, fear took hold:

> I have a spinal cord injury and thus impaired respiratory function. Which makes me basically have to hold my breath from the time I leave my car until I get into the school, since it otherwise feels like my lungs are filled with poisonous fumes.[21]

On the initiative of the municipal health authorities, a report was drawn up, which found that the smell of the encampment was insignificant in comparison to local emissions from industry and traffic. But the neighbours weren't having it and eventually – like a scene from the days of miasma – the police stormed in with one hundred men and cleared the area. In the aftermath, people showed no signs of regret: 'They should be thankful they haven't been shot . . . Here we have rules, and we Swedes follow them; these damned parasites on the Swedish society should fucking do what they're told . . . They only create problems for us honest Swedes.'[22]

The Smell of Holidays

Paradoxically, the decline in the diversity of smells can result in a sense of loss, even for Swedes. An odd kind of longing that can now only be relieved by a holiday – those brief, safe, obligation-free sorties out into the world with all our senses open wide. The spice-laden heat outside the airport shuttle in Mauritius, the humid stench of tens of thousands of mopeds in Hanoi, rotting fruit, hot leather and dried dog faeces in Old Cairo. All this can be inhaled in liberal doses without a glimmer of fear. Because soon we'll be going back home.

Nowadays, we have to travel to distant continents to experience exotic smells, but only a few decades ago, you didn't need to go any further than Italy, at least if you were a northern European. In Aldous Huxley's 1925 novel *Those Barren Leaves*, the proud owner of a grocery shop has managed to snare a British couple. The room stank of 'goat's milk cheese, pickled tunny, tomato preserve and highly flavoured sausage'.

> 'Whee-ew!' said Miss Thriplow, and pulling out a small handkerchief, she took refuge with the ghost of Parma violets . . .
>
> 'I forestieri sono troppo delicati. Troppo delicati' [Foreigners are too delicate. Too delicate], he repeated.
>
> 'He's quite right,' said Mr Cardan. 'We are. In the end, I believe, we shall have to sacrifice everything to comfort and cleanliness . . . As for this particular stink,' he sniffed the air, positively with relish, 'I don't really know what you have to object to it. It's wholesome, it's natural, it's tremendously historical. The shops of the Etruscan grocers, you may be sure, smelt just as this does. No, on the whole, I entirely agree with our friend here.'[23]

19

Hygienic Equivalents: The Smell of Nothing

Towards the end of the 1970s, growing numbers of people claimed they were becoming sick after spending time in modern buildings. Not just homes, but nurseries, schools and offices too. The symptoms included nausea, dizziness, itching, dry eyes, headaches, listlessness, concentration problems and an unusual increase in allergies and asthma.[1] This phenomenon, which was soon dubbed 'sick building syndrome', attracted considerable press coverage as sufferer after sufferer came forward, and the worst buildings were evacuated.

Scientists were bewildered. Regular air-quality monitoring equipment yielded no answers, and it proved impossible to find the cause.[2] But since many of those affected also complained of bad smells, odour reduction was quickly adopted as a strategy for dealing with the impasse.

Neo-Miasma

Like a kind of indestructible conspiracy theory, the perception of a link between smell and sickness still persists, not just among ordinary people but many health professionals.[3] And when scientists and engineers now set to work again, they were simply picking

VINYL

ODORANT	GROSS FORMULA	MOLECULAR MASS [U]	BOILING POINT [°C]	ODOUR DETECTION THRESHOLD [PPB]
Vinyl chloride	C_2H_3Cl	62	−13.4	10,000–20,000
Toluene	C_7H_8	92	111	500–5,000
Phenol	C_6H_6O	94	181	30–70
Ethylbenzene	C_8H_{10}	106	136	2,000–20,000
Methyl isobutyl ketone	$C_6H_{12}O$	100	117	100–880
Hydrochloric acid	HCl	36	85	275–1,000

Polyvinyl chloride plastic, PVC, was introduced in the 1920s. Among the world's most widely used plastics, it's found in everything from bank cards and garden hoses to furniture and toys. Unpacking a new PVC shower curtain is almost like finding yourself in a cross between a petrol station and a sweetshop. The room fills at once with an intensely sweet blend that has a chemical twist, and many will find that it irritates their eyes and makes their nostrils sting. The odour detection threshold for the substances it contains are high, so you will probably have ingested considerable doses – sometimes above the threshold beyond which they are harmful to health – before you have even smelt them.

up the baton dropped by the bewildered German physician Dr Pettenkofer a hundred years before. And they, too, soon stumbled over the same problems that had dogged scientists for a couple of thousand years. While many of the chemical components of the smells could, admittedly, be analysed and understood, the experience of smell itself and its impact on health could still be neither measured nor weighed.

And this is where a Danish civil engineer called Povl Ole Fanger, master of the ancient Sostrup Castle, comes in. He appears to have

In other plastics, the fully reacted molecule structures are usually too well bonded to be volatile. But unreacted remnants of a monomer called vinyl chloride will still evaporate from many PVC products. It smells sharp and sweet with a distinctly chemical tang, and the associated health risks include both liver damage and cancer.

In addition, all PVC products – whether flooring, wallcoverings, place mats, medical equipment or trampolines – will have had numerous ancillary materials added to ease production and improve usability. Several of these will also exude sweet-smelling substances, mainly toluene, which simply smells of sweet solvent, phenol, which also has a sweet and slightly tarry aroma, ethylbenzene, which stinks of petrol, and methyl isobutyl ketone, which smells of mothballs. In combination, they produce an intense super-sweet blend that is almost too much of a good thing. And there's good reason to take the warning seriously: these are substances that have major health effects. Unfortunately, some of the most harmful emissions – including those of added softening agents and flame retardants – don't smell at all.

When PVC products burn, the sharp chlorine-like stench of corrosive hydrochloric acid will be a crystal-clear warning sign. However, its high odour detection threshold will only be exceeded after the harmful effects have already set in. But the instant you become aware of it, you have no choice: run!

been a pleasant and honest philanthropist who, like Pettenkofer, was tireless in his efforts to improve people's living conditions. But within the rigid framework of technocracy, he nonetheless took an unorthodox approach. He proposed using the much-maligned, flesh-and-blood human nose as the instrument of measurement. He invented the odour panel.

The Odour Panel

Anyone who was allergic, a smoker or pregnant was winnowed out when setting up the well-trained panels, representative by both sex and age. They were trained in the same odour technique and were not permitted to suffer colds, use perfume or eat spicy food for the duration of the experiments. Nor were they allowed to talk to each other.[4]

The panel had to sniff their way through an enormous arsenal of odour samples, whose intensity was to be given in units known as olf. One olf was defined as the smell of a standard adult person whose skin had a total area of 1.8 square metres (19½ sq. ft) who worked in an office and had a so-called hygienic standard equivalent of 0.7 baths per day. In addition to intensity, the participants were also required to assess the qualities of the smells – first and foremost, how much of a nuisance they were perceived to be in different concentrations – measured in units called decipols.[5] The whole point was to establish a representative average for the population: a standard. If not at the global level, then at least for Western civilization.

Once this standard had been established, the odour nuisance in any indoor climate could be calculated, and it became possible to check it before and after implementing odour-reducing measures. Ole Fanger's method was soon at the very vanguard of the war against 'sick building syndrome'.

Odourless Poison

Naturally, objections were raised to the method, especially when it came to how representative it was.[6] After all, the perception of different smells varies greatly from one person to the next, often by a factor of several hundred for some substances. The same goes for the quality assessments. And it's not only that cultural background and life experience are absolutely crucial: particularly sensitive individuals may also find the scents of flowers and perfumes as much of a nuisance as sewage and old sweat.

At the same time, it soon became clear that several of the substances in the modern indoor climate that were most damaging to health didn't smell at all. Scientists had finally carried out more thorough studies of the emissions from the many new ultra-processed materials that were introduced in the wake of the hygiene hysteria of the 1950s, and had found respiratory irritants, allergens, mutagens and carcinogens. Where these had any odour at all, they tended to smell of candy and have an unspecified chemical odour that would not immediately alert people to the risk.

More Ventilation

The rational reaction to this new insight would, naturally, have been to take the harmful products out of circulation and replace them with tried and tested materials of solid wood, stone, ceramic and tile. But, of course, that idea jarred with a growth perspective in which an endless stream of new, highly industrialized products appeared to be the very driving force of the economy. Instead, a far less controversial strategy was adopted, an apparent answer to all the problems: boost the ventilation!

Since fan-based mechanical ventilation systems were first introduced in American skyscrapers in the 1940s, they have become obligatory in all buildings, even small homes. And over the course of the twenty-first century, building regulations throughout Europe have required ever-increasing volumes of air to be blasted through buildings. While this reduces the concentrations of the new materials' pollutants, of course, it also means that the building's own accumulated odours disappear and the scent signals between the people who pass through the rooms are short-circuited. Several layers of channel filters also ensure that any outside smells are eliminated. In modern schools, offices and public buildings, the olfactory silence is absolute.

Incubator

Of course, we can't pretend that Povl Ole Fanger and his colleagues weren't motivated by genuine altruism. After all, when smell vanishes, so too does part of the reason for stigmatizing women, peasants, manual workers and immigrants. While the slightly totalitarian effect was probably just an unintended consequence of standard technocratic rationality, it's just as absurd as making people wear ear defenders as long as there are still insulting words in the world or blinkers as long as there are still visible differences.

Research groups soon continued along the same lines, also establishing a standard for air temperature, where '0' on the scale represents 'I feel neither hot nor cold.' Once again, the system was tailored to the average population, even though we know that perceptions of temperature vary a great deal, not just between individuals, but between genders and age groups.[7]

Normality is defined as a sensory absence, then. While the ventilation systems maintain constant air exchange, the heating systems lock in the temperature – all of this, ideally, automated through a network of sensors. The user's own wishes and needs are overridden. In practice, we are in a state of incubation. The only thing that can rouse us is mechanical failure – when the thermostat breaks down and the fans collapse.

Triumph of the Eye

Our sensory apparatus is stimulated by variation and is sapped if unchallenged. As a child, Povl Ole's hands must almost certainly have felt the contrast between wet waterproof trousers and the warm, rough surface of the oak door that led into the Renaissance castle of Sostrup. And time after time, his body probably registered the transitions from dry and warm to cold and damp as he stumbled through the dark stone corridors. Above all, his nose must have picked up the shifting aromas from the fireplaces, the kitchens and the mouldy wine cellar. Perhaps he even detected the scent of his own family, and sporadic wafts of former lords of the castle.

Although Fanger and his associates were clearly unable to appreciate the qualities inherent in all this, it would perhaps be an exaggeration to assume that they weren't even capable of enjoying the chill breeze when the balcony doors were opened in the morning, the warmth of the sun on their forehead, and the scents of sudden summer rain and freshly ploughed fields. Even dogmatic technocrats are human after all.

Back in the 1960s, the social anthropologist Edward T. Hall already foresaw a civilization in which the sense of sight would gain dominion over all other sensory experiences.[8] A few decades later, we were already there, according to writer Italo Calvino: in 'an unending rainfall of images'.[9] Subjective experiences of quality, sensuality and temperament are replaced with standardized spreadsheets of scales, distances, perspectives and coordinates.[10] We have ended up in a pre-programmed world that is starting to look more and more like a mediocre computer game.

Temperature and Humidity

If it can be difficult to spot the psychological benefits of the prevailing ventilation strategy, it isn't much easier to identify any improvements in physical health. 'Sick building syndrome' still makes regular appearances in brand-new buildings built according to the very highest standards of hygiene. Even in super-ventilated health facilities, almost half of the workers may suffer from a poor indoor climate.[11] And, with the smells whisked away by ventilators, our noses are less capable than ever of alerting us to the sources or helping to track them down.

Despite all this, there is still some hope of recovering the sense of smell in our buildings – at least for the romantics among us. The first step is, of course, to reduce the need for high ventilation levels by eliminating ultra-processed materials, especially the plastic-based products. After that, we need to go back to John Scott Haldane in Chapter Twelve, whose snorkel experiment in the early twentieth century showed that it is, above all, the combination of high humidity and heat that causes problems with the indoor climates of our buildings.

Both things can largely be managed through architecture itself. Heavy constructions can stabilize temperatures within an acceptable interval, and porous materials like wood, tiles and lime-based plaster can absorb surplus humidity in the day and release it back into the rooms during the night. And perhaps buildings should have slightly higher ceilings to ensure that warm, humid exhalations have a temporary refuge on the busiest days until, again, they can be processed by the building materials themselves.[12] In a building constructed according to principles like these, our sense of smell can evolve freely, without obstacles or restrictions. And the pore-rich materials will simultaneously ensure that the building's entire olfactory history is recorded and stored.

Smell Silenced

In western Europe, we spend more than 90 per cent of our time indoors. From the moment we wake up in the morning in our warmish rooms to the hum of the ventilation system that has dealt with the emanations of the night. Then the lift down to the communal garage. The car's air-con kicks in automatically and escorts us through the rain to the garage at work. Up in another lift to rooms with smooth surfaces and, once again, the faint hum of the ventilation system – until the working day ends and we go back the same way we came, through the sluice system of garages, cars and lifts. Back home, everything is as it was when we left and we can confirm that the day has unfolded without contrasts or discomforts of any kind. And yes, this is a succession of clichés, but no less true for all that. It is a life in the spirit of Immanuel Kant, Sigmund Freud and Povl Ole Fanger.

But the rest of us know that this isn't enough. We are starting to look forward to the holidays again. The most desperate of us will opt to go so far away that we can hardly expect the air-con *not* to fail or the heating system *not* to break down – rousing the smell of ourselves, of others and of our surroundings to life once again.

20

The Scent Organ: The Smell that Remains

In *Brave New World*, Aldous Huxley's dystopian novel from 1932, the writer depicts an authoritarian society that is more like our own than we'd like to think. Images and screens are everywhere, and the cultivation of body and mind is spiralling to ever greater heights, driven by artificial reproduction and biotechnology. But above all else, this is a thoroughly sanitized world where 'civilization is sterilization.'[1] In order to preserve this ideal, excursions are organized to the so-called 'Savage Reservation', where Lenina – the main female character – constantly has to hold her nose. 'No air, no space; an under-sterilized prison; darkness, disease and smells.'[2]

The society that awaits them on their return home is not odourless. Far from it. But instead of the natural smells of the 'savages', they are met by a rich assortment of artificial scents, all of which perform a function. While some of them simply foster well-being, others are designed to prevent jealousy and stifle any inclination to pernicious erotic deviations. Nine different perfumes are delivered through the taps in the bathrooms, and there are public concerts on a scent organ.

> It was playing a delightfully refreshing Herbal Capriccio – rippling arpeggios of thyme and lavender, of rosemary, basil, myrtle, tarragon; a series of daring modulations through the

spice keys into ambergris; and a slow return through sandalwood, camphor, cedar and new-mown hay (with occasional subtle touches of discord – a whiff of kidney pudding, the finest suspicion of pig's dung) back to the simple aromatics with which the piece began.[3]

Synthetic Smells

Our own modern existence is hardly lacking in odours either. Although Povl Ole Fanger and the ventilation experts made a clean sweep in nurseries, schools and other public buildings, the world may never have been smellier. But the odours that now surround us are all manufactured and deployed with intention and ulterior motives – just like in Huxley's dystopia.

Things really took off in the latter half of the twentieth century, when the chemical industry became able to recreate simplified versions of some of the most attractive odour compounds.[4] And once the perfume industry had met all the demand for synthetic musk and floral scents, it set its sights on enhancing the cinema experience with technologies like Smell-O-Vision and AromaRama.[5] Different smells were released into the auditoriums via the ventilation system, although problems with precision meant that the public quickly ended up being assaulted by a blend of smells that did nothing but cause confusion. One of the more successful examples was John Water's 1981 gross-out comedy *Polyester* where audiences were given scratch-and-sniff cards with ten smells, from the whiff of cheesy socks, farts and pizza to the scent of new car and rose perfume. A flashing number on the screen told people when to scratch.

CHRISTMAS TREE

ODORANT	GROSS FORMULA	MOLECULAR MASS [U]	BOILING POINT [°C]	ODOUR DETECTION THRESHOLD [PPB]
Alpha-pinene	$C_{10}H_{16}$	136	156	0.05–7.3
Beta-pinene	$C_{10}H_{16}$	136	166	0.06–16
Limonene	$C_{10}H_{16}$	136	352	0.1–6
Bornyl acetate	$C_{12}H_{20}O_2$	196	227	0.01–4

Three terpenes dominate the more than fifty odorants that are produced by a newly felled spruce, pine or juniper Christmas tree: alpha-pinene, beta-pinene and limonene. These three compounds are extracted from the viscous resin and all share the same gross formula, which is based on carbon and hydrogen. However, owing to small variations in the shape of their atomic structures, alpha-pinene gives off a fresh, sweetish and slightly simple smell of turpentine and pinewood, while beta-pinene has a somewhat more complex and earthy scent, and limonene exudes an unmistakable aroma of citrus fruit. In addition to these resinous compounds, the tree's needles contribute a substance called bornyl acetate, which fills out the scent with a rich bouquet of camphor. All these substances have low odour detection thresholds, so only small doses are needed.

Natural variants that are produced by distilling resin – primarily from pine – are becoming increasingly rare. By contrast, enormous volumes of the synthetic versions are being manufactured and added to everything from perfumes for cleaning products to, of course, scent spray for synthetic Christmas trees. But even to an unsophisticated nose, this spray will seem strikingly one-dimensional compared with the complex natural version of a genuine Christmas tree, with its hints of soil, bark, vanilla, lilac and camphor, as well as a touch of sweat, cat pee and stinky dishcloths.

Soap and Freshly Baked Bread

Artificial smells can scare people or bring them together, inspire enthusiasm or anxiety. The modern fragrance industry doesn't just employ chemists, but psychologists and biologists too, whose job is to identify the optimal links between smells and emotions. The aim is to crack the behavioural code and, if possible, make the fake seem real, and the real seem even realer. There's no doubt about it: these are products of commercial interest.[6]

In the 1920s it was common to equip the lifts in American skyscrapers with light background music to calm the nerves of those using them for the first time.[7] When it turned out that music didn't just calm anxiety but boosted sales too, the retail sector followed suit. This was also the motivation for introducing new artificial smells, which were initially used in large shopping centres. Here, they were disseminated directly through the ventilation systems, which sent the safe, homey smells of new-baked bread and focaccia drifting through the glass corridors. Food shops emanated a light scent of lemons, soap stores the fragrance of exotic orchards, and cafés wafted aromas of chocolate and fresh coffee. The olfactory experience of those tempted in by the scents was immediately supplemented by warm vanilla and cinnamon, which have been shown to prolong the amount of time customers spend in stores, as well as increasing their willingness to buy.[8] Today, a shopping experience shaped by smells is the rule rather than the exception throughout the Western world, and we would be confused if it disappeared.

Christmas Spirit in a Can

Ever since the first detergents, soaps and deodorants were perfumed with floral scents, increasing numbers of consumer products have come to contain added synthetic odorants. Artificial aromas are added to food products, and fragrances are added to cosmetics, fabric softeners, toothpastes, textiles, erasers and toys. They often form the underlying smell of the modern home itself, with its combinations of lavender, rosemary, peppermint and coconut. Not to mention pine scent, which younger people will more readily associate with cleaning products than actual forests.

Except for those who still observe the tradition of a genuine Christmas tree, of course. But their numbers are steadily dwindling because it's far less effort to achieve Christmas spirit with a plastic Christmas tree drenched in synthetic pine-scented spray. Pay attention to the warning on the can though: 'Not for consumption. Avoid spraying on food.'[9] Why? Because every single one of the more than four hundred new industrial smells that are launched each year is an artificial chemical compound that may have very different health effects than its natural alternative, especially on people with allergies.[10] At the same time, their biodegradability is often poor and, as a result, they are accumulating in ever-increasing concentrations in humans, plants and animals. One example of this is the synthetic musk compounds that are used in massive quantities in everything from perfume to shampoo. They have now been detected in human blood, fatty tissue and breast milk, and are suspected of having oestrogen-like effects. The substances are also transmitted via sewerage systems to fish and marine molluscs. This has caused serious damage to the gill function of marine mussels.[11]

Manipulations

In principle, anything can be scented and a lot of things are – not just retail goods, either. The air in hotels, airport terminals, dental surgeries and hospitals is often treated with synthetic additives to camouflage bad smells or counteract anxiety, for example. Smell has also been shown to prevent claustrophobia in cramped settings, and the smell of a log fire appears to reduce stress in intimidatingly large rooms.[12] The intentions may be benign, but these are all calculated lies – regardless of how white they are.

The deception takes on a more brutal dimension when, say, jasmine is added to the air in an office to reduce the number of keystroke errors among data operators, or bills are sprinkled with musk to encourage prompt payment, or meeting rooms are sprayed with testosterone to heighten conflict levels, or with congenial floral scents to soothe tempers and increase tolerance.[13]

The artificial fragrances used to deodorize modern society 'are evocative of things which are not there, of presences which are absent' according to the Canadian cultural historian Constance Classen. They 'are a sign without a referent, smoke without fire, pure olfactory image'.[14] And when the signs are fake, our understanding of reality breaks down.

The Smells that Are Left

The Western world's neglect of the sense of smell is likely the reason why we unhesitatingly accept plastic Christmas trees, vanilla-free vanilla ice cream, smoked ham that has never been smoked, wooden floors without a trace of timber, and the idea of reducing the smell of our fellow humans to sham scents of flower meadows and arousal.

To all intents and purposes, authentic and complex smells are suppressed and absent from our existence, other than the sporadic occasions when we still cook with natural raw ingredients. We can, of course, go out into nature and sniff around for aromatic signals of the symbiosis between fungi and plants, or between plants and animals when flowers unfurl and fruit ripens.

Or we can go to the library.

The Smell of Books

It has to be one of those old libraries with dusty ceiling lights, floor-to-ceiling shelves, discreet Danish-speaking staff and a slight, sharp whiff of mites. Books are never washed and the porous paper in the oldest tomes has stood around gathering odours for hundreds of years – from the inner life of the library and from the borrowers who have read their way through the pages in sorrow or delight.

Smelling a new book is a quick business. If the pages are porcelain-white and glossy, generous doses of glue, synthetic rubber and bleach will ensure a sharp, acidic stench of motor oil, which drowns out any other nuances. Books made with ordinary paper have a little more to offer, with their faint whiff of fresh cellulose and vinegar, and maybe some undertones of spices from the soy-based printing ink. The cheapest paperbacks are slightly more complex, and you may sometimes pick up dark tones with a hint of rotten egg and cowpat. But that's the exception.

The true olfactory experiences lie concealed in the old leather-bound books. And you most definitely need to borrow them and take them home so that you can be sure of conducting your experiments in the requisite peace and quiet. It doesn't matter what the book is about – you aren't going to read it. But remember that it must be well used. And it's best to keep it closed overnight until it has warmed to room temperature. The next morning, you'll find that the room is filled with a faint tang of leather, fur and birch oil. Now, open the book quickly at a random page and draw a deep breath. Let the book lie open for a few seconds afterwards to allow the smells to develop further, then sniff again. If you're in luck, your nose will fill with warm vanilla, sawdust, camphor and chocolate, often with a hint of black soil. Another book may give off smoke, coffee, walnut and clay, and a third, flowers and sweat. If the book has been kept in a damp storeroom: marzipan, fungus and soggy earth. It varies a great deal. With a little extra concentration and imagination, you may even catch the tiniest waft of human on the outer edge of the pages when you warm them between your fingers and scrape them gently with your nail.

When Sight Vanished

In his 1995 novel *Blindness*, the Portuguese writer José Saramago conducts a dramatic thought experiment. He imagines a population struck blind by an epidemic. Suddenly, all anyone can see is white.

The city soon becomes a chaotic jungle of terrified people, roaming around with their arms stretched out in front of them. Their world is one that is based entirely on sight. The other senses are untrained and incapable of compensating for this loss. Within days the authorities' capacity to run the city and exercise power has collapsed. After that, all criminality, corruption and violence stop too. Without the gaze and its capacity to inculcate shame, taboos also vanish, one after another.

'With all of us ending up blind, as appears to be happening, who's interested in aesthetics, and as for hygiene, tell me, doctor, what kind of hygiene could you hope for in this place,' wonders a troubled patient. 'Perhaps only in a world of the blind will things truly be what they are,' replies the doctor.[15]

Eventually the citizens – now a touching, tasting, listening and smelling multitude – seek each other out, forming groups in order to survive; life gradually calms down and even becomes pretty enjoyable – until, one day, their sense of sight suddenly returns.

Afterword

In a senseless age where folly, bluff and propaganda are encroaching on more and more of our life, it may be some small consolation to find the tiny cracks where authentic smells still leak out. Then again, why not resort to more proactive measures?

Like both the Futurists and the Decadents before them, some modern artists have defied the spirit of the age. The exhibitions of Norway's Sissel Tolaas offer us odours such as old trains, dirty banknotes and stubborn stress sweat, not to mention bars of cream soap infused with the scents of the artist's own body.[1] These are beautiful and interesting provocations, but only available to audiences with an interest in art. The question is, can we also take action in everyday life?

Let me start off by saying that regular handwashing is rarely a bad idea, especially when pandemics are raging. That said, you have little to lose by dropping deodorant and cutting down on showers – or at least restricting your cleaning efforts to your bodily orifices. This could even improve your immune system, and your lush sweaty emanations will guarantee you plenty of elbow room on public transport. The downside is that it may also cost you friends and colleagues. If you up your intake of cabbage and beans instead, you can always ensure a little extra flatulence in the workplace and at social events, although you'll risk your job and your chances of future invitations.

Perhaps, in sheer frustration, you'll take to the streets to conduct a systematic campaign of public urination, but the hefty fines you'll incur for anti-social behaviour will soon leave you bankrupt.

The situation will quickly start to look pretty hopeless, but if you're a truly desperate soul, there is still one other option: the stink bomb. It's a method that lends itself to swift and efficient olfactory statements in high streets and shopping centres, schools and government canteens. And you can load the stink bombs with a variety of odours – 'good' or 'bad', depending on your intent.

Unfortunately, supermarkets, restaurants and hotels already have a monopoly on the stunted versions of the 'best' scents, so only 'bad' odours have enough of a kick to rouse our slumbering sense of smell. That being the case, it seems only natural to seek inspiration from an authority in the field: the U.S. military.

Back in the early 2000s, they decided to develop the ultimate stink bomb. They outsourced the project to Monell Chemical Senses Center in Philadelphia, whose job it was to design a smell so unpleasant that enemy troops would not just flee the battlefield but end up in a state of prolonged fear and nausea. Unsurprisingly, a smell known as 'U.S. Government Standard Bathroom Malodor' was selected as the key ingredient. A blend of eight different chemicals, it smelt like faeces, only stronger. When they supplemented this with a sulphur-based component based on rotting animal carcasses, the scientists were certain they'd found a sure-fire winner. But early in the testing phase the blend quickly proved to be far less effective than expected. And once they realized that it was impossible to find a smell that was unequivocally repulsive across all cultures, the project was shelved.[2] Perhaps it would have helped if they'd supplemented it with a notorious stench that the Boxers resorted to during their rebellion in late nineteenth-century China: the binding cloths from women's feet.[3] But it isn't very likely.

Like everything else to do with smell, it's best to think locally. And in case you also wish to act locally, I will close with a recipe that is pretty certain to blow the noses off most Europeans and North Americans – we have, after all, become so homogeneous. It is, of course, based on natural and completely biodegradable ingredients and is prepared as follows:

> Crack an egg into an empty bottle. Add one dessert spoon of milk and one of household vinegar. Screw on the top and leave the bottle in a warm spot, preferably in the sun. After a single day, you will already be able to smell it, and the stink will rapidly become extremely pungent. That's when it's time to deploy it outdoors, preferably in a rubbish bin or, more anonymously, along with a bunch of other empties. I guarantee you it will beat even the most effective ventilation system.

But no, dear reader. Don't pay me any mind. I'm just hyperventilating.

REFERENCES

Introduction

1 Nikos Kazantzakis, *Report to Greco*, tr. P. A. Bien (Simon & Schuster, 1965).

2 Idan Frumin et al., 'A Social Chemosignaling Function for Human Handshaking', *eLife*, 4 (2015), art. e05154.

3 Jean-François Lyotard, *L'Inhumain: Causeries sur le temps* [The Inhuman: Conversations about Time] (Éditions Galilée, 1988).

4 Helene Asphaug, 'Stivpynta feminister. Du er ikke frigjort når du bruker mer tid på å bli sett enn hørt' [You Are Not Liberated if You Spend More Time on Being Seen than Heard], *NRK Ytring*, 23 April 2022.

5 Juhani Pallasmaa, *The Eyes of the Skin: Architecture and the Senses* (John Wiley & Sons Ltd, 2005).

6 McCann Worldgroup survey referred to in A. S. Barwich, *Smellosophy: What the Nose Tells the Mind* (Harvard University Press, 2020).

7 Pallasmaa, *The Eyes of the Skin*, p. 32.

8 David Michael Levin, ed., *Modernity and the Hegemony of Vision* (University of California Press, 1993).

9 Matthew Cobb, *Smell: A Very Short Introduction* (Oxford University Press, 2020).

10 Ibid.

11 Barwich, *Smellosophy*.

12 Günther Ohloff et al., *Scent and Chemistry: The Molecular World of Odors*, 2nd edn (Wiley-VCH, 2022).

13 The information in this and subsequent tables is drawn from research papers by Ohloff et al., ibid., that are publicly available on Google Scholar and various Internet databases, including PubChem

(National Centre for Biotechnology Information). The figures cited must be considered approximate since they may be calculated using different methodology.

14 Avery N. Gilbert and Charles J. Wysocki, 'The Smell Survey Results', *National Geographic*, 172 (October 1987), pp. 514–25.

15 Ohloff et al., *Scent and Chemistry*, p. 31.

16 Cobb, *Smell*.

17 This information was sourced from, among others, Margit Walsø, *Din femte sans. Hvordan luktesansen beskriver deg, din fortid og din fremtid* [Your Fifth Sense: How Your Sense of Smell Describes You, Your Past and Your Future] (Kagge forlag, 2022).

18 Gas chromatography and mass spectrometry are often used in combination to analyse gases, but they play different roles. The first separates out the various substances based on their movement through the gas phase and absorption in a material, whereas mass spectrometry measures the mass of ionized particles to identify and quantify molecules in the sample. In other words, gas chromatography helps to sort the substances, whereas mass spectrometry provides information about the components of each individual substance.

19 Trygg Engen, *The Perception of Odors* (Academic Press, 1982).

20 Ibid.

21 Rachel S. Herz, 'I Know What I Like: Understanding Odor Preferences', in *The Smell Culture Reader*, ed. Jim Drobnick (Berg, 2006), pp. 190–203.

22 Ibid.

23 Robert Muchembled, *Smells: A Cultural History of Odours in Early Modern Times*, tr. Susan Pickford (Polity, 2020).

24 A. Arshamian et al., 'The Perception of Odor Pleasantness Is Shared Across Cultures', *Current Biology*, 32/9 (2022), pp. 2061–6.

25 Gilbert and Wysocki, 'Smell Survey'.

26 Åke Bresle, *Näsa för lukter* [A Nose for Smells] (Ingejörsförlag, 1986).

27 Gemma Francis, 'Freshly Baked Bread Tops Poll of Britain's Top 50 Favourite Smells', *Daily Mirror*, www.mirror.co.uk, 25 May 2015.

28 Alan R. Hirsch, 'Nostalgia, the Odors of Childhood and Society', in *The Smell Culture Reader*, ed. Drobnick, pp. 187–9.

29 Cretien van Campen, *The Proust Effect: The Senses as Doorways to Lost Memories*, tr. Julian Ross (Oxford University Press, 2014).

30 Engen, *The Perception of Odors.*
31 Marcel Proust, *In Search of Lost Time* [1913], tr. C. K. Scott Moncrieff and Terence Kilmartin, rev. D. J. Enright (The Modern Library, 2003).
32 Hirsch, 'Nostalgia'.
33 Ali Lakhani et al., 'Using the Natural Environment to Address the Psychosocial Impact of Neurological Disability: A Systematic Review', *Health and Place*, 55 (January 2019), pp. 188–201.
34 Kei M. Igarashi et al., 'Coordination of Entorhinal–Hippocampal Ensemble Activity During Associative Learning', *Nature*, 510 (2014), pp. 143–7.
35 Cobb, *Smell.*
36 Jon Fosse, 'Eg saknar Fornebu' [I Miss Fornebu], in *Luft og kjærlighet* [Air and Love], ed. V. Jordal et al. (Dinamo forlag, 2008), p. 227.
37 Angelique Chrisafis, 'Proust's Memory-Laden Madeleine Cakes Started Life as Toast, Manuscripts Reveal', *The Guardian*, www.theguardian.com, 19 October 2015.
38 Henriette Schønberg Erken, *Stor kokebok for større og mindre husholdninger* [Big Cookbook for Households Large and Small, 1914] (H. Aschehoug & Co., 1941).

1 The Midden: The Smell of Safety

1 Karl Ove Knausgård, *A Death in the Family: My Struggle Book 1* [2009], tr. Don Bartlett (Harvill Secker, 2012).
2 Halldór Laxness, *Independent People*, tr. J. A. Thompson (Vintage Classics, 2008).
3 Arne Lie Christensen, *Den norske byggeskikken. Hus og bolig på landsbygda fra middelalderen til vår egen tid* [Norwegian Building Traditions: Houses and Homes in Villages from the Middle Ages to Our Own Times] (Pax forlag, 1995).
4 Cited ibid.
5 George Steuart Mackenzie, *Travels in the Island of Iceland, During the Summer of the Year MDCCCX* (Archibald Constable and Company, 1811).
6 Laxness, *Independent People.*
7 Gaston Bachelard, *The Poetics of Space* [1964], tr. Maria Jolas (Beacon Press, 1969).

8 Knausgård, *A Death in the Family*.
9 Ibid.
10 Bjørn Berge, *The Ecology of Building Materials*, 2nd edn, tr. Chris Butters and Filip Henley (Taylor & Francis Group, 2009).
11 Rainer Maria Rilke, *The Notebooks of Malte Laurids Brigge* [1910], tr. Stephen Mitchell (Vintage Books, 1989).
12 Rachel Ehrenberg, 'Urban Microbes Come Out of the Shadows', *Nature*, 522/7557 (2015), pp. 399–400.
13 Sveinung Bang-Andersen, *Svarthåla på Viste – boplass i 6000 år* [The Black Cave – A Stone Age Dwelling Place] (AmS-Småtrykk, 1983).
14 Rachel Herz, *That's Disgusting: Unraveling the Mysteries of Repulsion* (W. W. Norton & Company, 2012).

2 A Feast for the Senses: The Smell of Pleasure

1 Rodolfo Lanciani, *Ancient Rome in the Light of Recent Discoveries*, 12th edn (Houghton Mifflin & Company, 1898), p. 65.
2 Neville Morley, 'Urban Smells and Roman Noses', in *Smell and History: A Reader*, ed. Mark M. Smith (West Virginia University Press, 2019), pp. 33–49 (p. 34).
3 Gemma Jansen, 'Sewers or Cesspits? Modern Assumptions and Roman Preferences', in *Latrinae: Roman Toilets in the Northwestern Provinces of the Roman Empire*, ed. S. Hoss, Roman Archaeology Series (Archaeopress, 2018), pp. 5–18.
4 Ibid.
5 Calculated on the basis of 2 litres of urine and 0.4 kilograms of excrement per person per day.
6 Morley, 'Urban Smells', p. 43.
7 Excerpt from Juvenal's third Satire from around 110 CE. Cited in Annie Riis, *Satirens tid. Dikt og gjendiktninger* [The Age of Satire: Poems and Translations] (Aschehoug, 1994), p. 30.
8 Ann Olga Koloski-Ostrow, 'Roman Urban Smells: The Archaeological Evidence', in *Smell and the Ancient Senses*, ed. Mark Bradley (Routledge, 2015), pp. 90–109.
9 Morley, 'Urban Smells', p. 44.
10 Ibid.
11 James K. Moran et al., 'The Scent of Blood: A Driver of Human Behavior?' *PLOS One*, 10/9 (23 September 2015), at https://journals.plos.org.

12 Sveinung Berg Bentzrød, 'De norske Kabul-historiene' [The Norwegian Kabul Stories], *Aftenposten*, www.aftenposten.no, 3 September 2021.
13 Roland Auguet, *Cruelty and Civilization: The Roman Games* (Routledge, 1994).
14 Constance Classen, David Howes and Anthony Synnott, *Aroma: The Cultural History of Smell* (Routledge, 1994), p. 26.
15 Ibid.
16 Jonathan Reinarz, *Past Scents: Historical Perspectives on Smell* (University of Illinois Press, 2014).
17 Classen, Howes and Synnott, *Aroma*, p. 21.
18 Morley, 'Urban Smells', p. 111.
19 Allison Emmerson, *Life and Death in the Roman Suburb* (Oxford University Press, 2020).
20 Ibid.

3 Heavenly Pastures: The Smell of Salvation

1 Charles Dickens, *The Uncommercial Traveller* (Chapman and Hall, 1905), p. 89.
2 Dan McKenzie, *Aromatics and the Soul: A Study of Smells* (William Heinemann, 1923).
3 Yngve Vogt, 'Reformasjonen tok flere hundre år' [The Reformation Took Several Hundred Years], *Apollon*, 29/2 (April 2019), at www.uio.no/forskning.
4 Constance Classen, *The Color of Angels: Cosmology, Gender and the Aesthetic Imagination* (Routledge, 1998), p. 55.
5 Constance Classen, 'The Breath of God: Sacred Histories of Scent', in *The Smell Culture Reader*, ed. Jim Drobnick (Berg, 2006), pp. 375–90 (p. 381).
6 Günther Ohloff et al., *Scent and Chemistry: The Molecular World of Odors*, 2nd edn (Wiley-VCH, 2022), p. 2.
7 All bible quotations are taken from the King James Bible.
8 Cited by Ashley Clements in 'Divine Scents and Presence', in *Smell and the Ancient Senses*, ed. Mark Bradley (Routledge, 2015), pp. 46–59.
9 Jerry Toner, 'Smell and Christianity', in *Smell and the Ancient Senses*, ed. Bradley, pp. 158–70.

10 Cited in Constance Classen, David Howes and Anthony Synnott, *Aroma: The Cultural History of Smell* (Routledge, 1994), p. 51.
11 Issei Yokoyama et al., 'Inhalation of Odors Containing DMHF Generated by the Maillard Reaction Affects Physiological Parameters in Rats', *Scientific Reports*, 10 (2020), art. 13931, at www.nature.com.
12 Classen, *The Color of Angels*, p. 59.
13 Ibid.
14 Fyodor Dostoevsky, *The Karamazov Brothers* [1880], tr. Constance Garnett (Wordsworth Editions Limited, 2007).
15 Classen, 'The Breath of God', p. 376.
16 Ibid.
17 Dostoevsky, *The Karamazov Brothers*.
18 Ibid.
19 Ibid.
20 R.E.D. Clark, 'The Spheres of Revelation and Science: What Are Their Limitations in Relation to Each Other?', *Journal of the American Scientific Affiliation*, 5 (June 1953), pp. 8–17.
21 John Last, 'The Centuries-Long Quest for The Scent of God', *Noēma*, www.noemamag.com, 28 July 2022.

4 A Man's World: The Smell of Condemnation

1 Miguel de Cervantes Saavedra, *Don Quixote* [1605; 1615], tr. James H. Montgomery (Hackett Publishing Company, 2009).
2 Laura Alho et al., 'Nosewitness Identification: Effects of Negative Emotion', *PLOS One*, 10/1 (22 January 2015), at https://journals.plos.org.
3 Kate Fox, *The Smell Report: An Overview of Facts and Findings* (Social Issues Research Centre, 1999).
4 Patrick Süskind, *Perfume: The Story of a Murderer*, tr. John E. Woods (Penguin Essentials, 2015).
5 Cited in Constance Classen, David Howes and Anthony Synnott, *Aroma: The Cultural History of Smell* (Routledge, 1994), p. 30.
6 Robert Muchembled, *Smells: A Cultural History of Odours in Early Modern Times*, tr. Susan Pickford (Polity, 2020), p. 37.
7 Ibid., p. 85.
8 Ibid., p. 66.

9 C. M. Woolgar, 'Medieval Smellscapes', in *Smell and History: A Reader*, ed. Mark M. Smith (West Virginia University Press, 2019), pp. 50–75 (p. 59).
10 Nir Sterer and Mel Rosenberg, *Breath Odors: Origin, Diagnosis, and Management* (Springer, 2011).
11 Cited in Classen, Howes and Synnott, *Aroma*, p. 30.
12 Muchembled, *Smells*, p. 67.
13 Jonathan Reinarz, *Past Scents: Historical Perspectives on Smell* (University of Illinois Press, 2014).
14 Constance Classen, 'The Witch's Senses: Sensory Ideologies and Transgressive Femininities from the Renaissance to Modernity', in *Empire of the Senses: The Sensual Culture Reader*, ed. David Howes (Routledge, 2004), pp. 70–84.
15 Constance Classen, 'The Odor of the Other: Olfactory Symbolism and Cultural Categories', *Ethos*, 20/2 (1992), p. 144.
16 Cited in Muchembled, *Smells*, p. 87.
17 Rune Blix Hagen, *Hekser. Fra forfølgelse til fortryllelse* [Witches: From Persecution to Enchantment] (Humanist forlag, 2010).

5 The Devil's Breath: The Smell of Corruption

1 James Joyce, *A Portrait of the Artist as a Young Man* [1916] (Wordsworth Editions, 1992).
2 Ibid.
3 Hildegard of Bingen, *The Book of the Rewards of Life*, tr. Bruce W. Hozeski (Oxford University Press, 1997).
4 Jonathan Reinarz, *Past Scents: Historical Perspectives on Smell* (University of Illinois Press, 2014), p. 43.
5 Per Einar Odden, *Den hellige Kristina den Forbløffende (av Belgia), (1150–1224)* [St Christina the Astonishing of Belgium], Catholic Church of Norway, www.katolsk.no, 16 February 2001.
6 Constance Classen, 'The Breath of God: Sacred Histories of Scent', in *The Smell Culture Reader*, ed. Jim Drobnick (Berg, 2006), pp. 375–90 (p. 383).
7 Robert Muchembled, *Smells: A Cultural History of Odours in Early Modern Times*, tr. Susan Pickford (Polity, 2020), p. 17.
8 Constance Classen, David Howes and Anthony Synnott, *Aroma: The Cultural History of Smell* (Routledge, 1994), p. 55.
9 Muchembled, *Smells*, p. 22.

10 Dolly Jørgensen, 'Modernity and Medieval Muck', *Nature and Culture*, 9/3 (2014), pp. 225–37.
11 Harald Bjørkvik, *Byar og bysamfunn. Om hvordan byar er blitt til* [Cities and City Societies: On How Cities Came into Being] (Lillesandmuseet, 2017).
12 Jørgensen, 'Modernity and Medieval Muck'.
13 Ibid.
14 Norbert Elias, *The Civilizing Process: The History of Manners and State Formation and Civilization* [1939] (Blackwell, 1969). First published in English in 1969.
15 Marieke Lucas Rijneveld, *The Discomfort of Evening*, tr. Michele Hutchison (Faber & Faber, 2020).
16 Alain Corbin, *The Foul and the Fragrant: Odor and the French Social Imagination* (Harvard University Press, 1986).
17 Luther cited in Classen, Howes and Synnott, *Aroma*, p. 59.
18 Muchembled, *Smells*, p. 91.
19 Ibid.
20 Joseph P. Byrne, *Daily Life During the Black Death* (Greenwood, 2006), p. 132.
21 Daniel Defoe, *A Journal of the Plague Year* (Longmans, Green and Co., 1896).
22 Muchembled, *Smells*, p. 91.
23 Classen, Howes and Synnott, *Aroma*, p. 60.
24 Muchembled, *Smells*, p. 102.
25 Ibid., p. 104.
26 Byrne, *Daily Life*, p. 49.
27 Muchembled, *Smells*, p. 60.
28 Defoe, *A Journal of the Plague Year*, p. 243.
29 Ibid., p. 266.
30 Muchembled, *Smells*, p. 95.
31 Ibid., p. 104.

6 The Musk Deer: The Smell of Deception

1 Ji Qing-hong, 'Marco Polo and Musk: On the Authenticity of Marco Polo's Visit to China', *China Tibetology*, 1 (2020), pp. 60–67.
2 Anya H. King, *Scent from the Garden of Paradise: Musk and the Medieval Islamic World*, Islamic History and Civilization series (Brill, 2017), p. 360.

3 Ji Qing-hong, 'Marco Polo and Musk'.
4 Ibid.
5 Robert Muchembled, *Smells: A Cultural History of Odours in Early Modern Times*, tr. Susan Pickford (Polity, 2020), p. 132.
6 Günther Ohloff et al., *Scent and Chemistry: The Molecular World of Odors*, 2nd edn (Wiley-VCH, 2022), p. 333.
7 Bjørn Bandlien, 'Bever med baller' [Beavers with Balls], *Norsk Folkeminnelag*, 58 (2011), pp. 20–23.
8 Ohloff et al., *Scent and Chemistry*, p. 333.
9 Alan Hyde, 'Offensive Bodies', in *The Smell Culture Reader*, ed. Jim Drobnick (Berg, 2006), pp. 53–8 (p. 54).
10 Muchembled, *Smells*, p. 11.
11 Constance Classen, David Howes and Anthony Synnott, *Aroma: The Cultural History of Smell* (Routledge, 1994), p. 70.
12 Ibid.
13 Cited in Muchembled, *Smells*, p. 69.
14 Classen, Howes and Synnott, *Aroma*, p. 71.
15 Muchembled, *Smells*, p. 137.
16 Ibid., p. 140.
17 According to the English Wikipedia page, 'Eau de Cologne'.
18 Cited in F. P. Wilson, ed., *The Plague Pamphlets of Thomas Dekker* (Clarendon Press, 1925).
19 Classen, Howes and Synnott, *Aroma*, p. 50.
20 Cited ibid., p. 33.
21 Ohloff et al., *Scent and Chemistry*, p. 106.
22 Constance Classen, *The Color of Angels: Cosmology, Gender and the Aesthetic Imagination* (Routledge, 1998), p. 151.

7 The Liberated Gut: The Smell of Happiness

1 C. M. Woolgar, 'Medieval Smellscapes', in *Smell and History: A Reader*, ed. Mark M. Smith (West Virginia University Press, 2019), pp. 50–75 (p. 53).
2 Dante Alighieri, *The Divine Comedy*, tr. Henry Wadsworth Longfellow (Houghton, Mifflin and Company, 1867), p. 68.
3 Kari Ellen Gade, 'Einarr Þambarskelfir's Last Shot', *Scandinavian Studies*, 67/2 (1995), pp. 153–62.
4 Cited in Robert Muchembled, *Smells: A Cultural History of Odours in Early Modern Times*, tr. Susan Pickford (Polity, 2020), p. 39.

5 Rachel Herz, *The Scent of Desire: Discovering Our Enigmatic Sense of Smell* (William Morrow, 2007).
6 Muchembled, *Smells*, p. 60.
7 William Ian Miller, *The Anatomy of Disgust* (Harvard University Press, 1997), pp. 55–6. The faeces were artificial!
8 Muchembled, *Smells*, p. 9.
9 Ibid.
10 Ibid., p. 29.
11 Jae Kwak et al., 'Changes in Volatile Compounds of Human Urine as It Ages: Their Interaction with Water', *Journal of Chromatography B*, 941 (2013), pp. 50–53.
12 Peter Englund, *Fortidens landskap. Om mordet på Karl XII og andre essays* [The Landscapes of the Past: On the Murder of Charles XII and Other Essays], tr. (into Norwegian) Per Qvale (Universitets forlag, 1992), p. 148.
13 Jonathan Reinarz, *Past Scents: Historical Perspectives on Smell* (University of Illinois Press, 2014).
14 Cited in Muchembled, *Smells*, pp. 59–60.
15 Michel Foucault, *The History of Sexuality*, vol. I: *An Introduction*, tr. Robert Hurley (Pantheon Books, 1978).
16 Dominique Laporte, *History of Shit* [1978], tr. Nadia Benabid and Rodolphe El-Khoury (MIT Press, 2002).
17 Muchembled, *Smells*, p. 60.
18 Johan Vogt, *33 Regler for god tone på Grini* [33 Rules for Good Form at Grini] (Aschehoug, 1945).
19 Thomas Fuchs, *A Concise Biography of Adolf Hitler* (Berkley Publishing Group, 2000).
20 Cited in Sean Adams, 'John Cleese on Dark Comedy, His Best Monty Python Moment and His Upcoming Show at Hershey', www.pennlive.com, 7 October 2021; see, among others, *Monty Python and the Holy Grail* (1975) and *Monty Python's Flying Circus* (1969–74).
21 See, among others, Louis de Funès in *La soupe aux choux* [Cabbage Soup] (1981), Leslie Nielsen in *The Naked Gun* (1988), Jim Carrey in *Dumb and Dumber* (1994) and Eugene Levy in *The Man* (2005).
22 Jean Nohain and François Caradec, *Le Pétomane: 1857–1945* (Sherbourne Press, Inc., 1967).

8 Enter the Eye-Man: The Smell of Reason

1 Gordon Campbell, *The Hermit in the Garden: From Imperial Rome to Ornamental Gnome* (Oxford University Press, 2013), p. 22.
2 Ibid.
3 Oluwatosin Abdulsalam et al., 'Phytohormones and Volatile Organic Compounds, Like Geosmin, in the Ectomycorrhiza of *Tricholoma vaccinum* and Norway Spruce (*Picea abies*)', *Mycorrhiza*, 31 (2021), pp. 173–88.
4 Alain Corbin, 'The New Calculus of Olfactory Pleasure', in *The Smell Culture Reader*, ed. Jim Drobnick (Berg, 2006), pp. 167–80 (p. 175).
5 Robert Muchembled, *Smells: A Cultural History of Odours in Early Modern Times*, tr. Susan Pickford (Polity, 2020), p. 33.
6 Jean-Jacques Rousseau, *Of the Social Contract and Other Political Writings*, tr. Quintin Hoare (Penguin Books, 2012), p. 10.
7 Jonathan Reinarz, *Past Scents: Historical Perspectives on Smell* (University of Illinois Press, 2014), p. 13.
8 Alain Corbin, 'Why Smell the Past?', in *Smell and History: A Reader*, ed. Mark M. Smith (West Virginia University Press, 2019), pp. 1–12 (p. 9).
9 Cited in Dan McKenzie, *Aromatics and the Soul: A Study of Smells* (William Heinemann, 1923), p. 95.
10 Immanuel Kant, *Anthropology from a Pragmatic Point of View* [1798], tr. Mary J. Gregor (Martinus Nijhoff, 1974).
11 Kate Fox, *The Smell Report: An Overview of Facts and Findings* (Social Issues Research Centre, 1999).
12 Han Baltussen, 'Ancient Philosophers on the Sense of Smell', in *Smell and the Ancient Senses*, ed. Mark Bradley (Routledge, 2015), pp. 30–45 (p. 30).
13 Ibid., p. 45.
14 Nils-Gunnar Toremalm, *En doft av Linné* [A Scent of Linnaeus] (Sällskapet Heimdall, 2020).
15 Corbin, 'Why Smell the Past?'.
16 David Howes, 'The Craft of the Senses', Concordia University Occasional Paper, 2011.
17 Jon Røyne Kyllingstad, *Rase. En vitenskapshistorie* [Race: A Scientific History] (Cappelen Damm, 2023).
18 Constance Classen, cited in David Howes, 'The Social Life of the Senses,' *Ars Vivendi Journal*, 3 (February 2013), pp. 4–23.

9 Soda and Sulphuric Acid: The Smell of Prosperity

1 Nikolai Gogol, *The Nose* [1836], tr. Ronald Wilks (Penguin Classics, 2015), p. 12.
2 Ibid., p. 28.
3 Jonathan Reinarz, *Past Scents: Historical Perspectives on Smell* (University of Illinois Press, 2014), p. 172.
4 Michael Donovan, *Domestic Economy*, vol. 1: *Brewing, Distilling, Wine-Making, Baking &c.* (Longman, Rees, Orme, Brown and Green, 1830).
5 R. M. Martin, *The Past and Present State of the Tea Trade of England, and of the Continents of Europe and America* (Parbury, Allen & Co., 1832), p. 115.
6 Langdon Winner, *The Whale and the Reactor* (The University of Chicago Press, 1986), p. 24.
7 Hugh Miller, *First Impressions of England and Its People* (William P. Nimmo, 1869), p. 208.
8 Newcastle upon Tyne Town Council, *Proceedings and Reports of the Town Council of the Borough of Newcastle for 1839*, p. 19.
9 Alain Corbin, *The Foul and the Fragrant: Odor and the French Social Imagination* (Harvard University Press, 1986), p. 129; Peter Thorsheim, *Inventing Pollution: Coal, Smoke, and Culture in Britain since 1800* (Ohio University Press, 2006).
10 Report from Institut Chemists, 1809, cited in Corbin, *The Foul and the Fragrant*, p. 132.
11 Thorsheim, *Inventing Pollution*.
12 Hans J. Rindisbacher, *The Smell of Books: A Cultural-Historical Study of Olfactory Perception in Literature* (University of Michigan Press, 1992), p. 104.
13 Margit Walsø, *Din femte sans. Hvordan luktesansen beskriver deg, din fortid og din fremtid* [Your Fifth Sense: How Your Sense of Smell Describes You, Your Past and Your Future] (Kagge forlag, 2022), p. 52.
14 Ibid.
15 Klima- og forurensningsdirektoratet, *Regulering av luktutslipp i tillatelser etter forurensningsloven* [Regulation of Odour Emissions in Permits as per the Pollution Control Act] (2013), p. 8.
16 Michael S. Carolan, 'When Good Smells Go Bad: A Sociohistorical Understanding of Agricultural Odor Pollution', *Environment and Planning A: Economy and Space*, 40/5 (2008), pp. 1235–49.

17 Benjamin Fredriksen, 'Jobben til Ansgar er å lukte på tørrfisk: – Det lukter gull' [Ansgar's Job Is to Sniff Dried Fish: It Smells of Gold], NRK *Nordland*, www.nrk.no, 30 August 2015.

10 The Sweat of the Earth: The Smell of Fear

1 Alain Corbin, *The Foul and the Fragrant: Odor and the French Social Imagination* (Harvard University Press, 1986).
2 Report by J-N Hallé, cited ibid., p. 2.
3 Corbin, *The Foul and the Fragrant*, p. 4.
4 Ibid., p. 15.
5 The Dutch physician, Picter van Musschenbroek, cited ibid., p. 22.
6 Report from the Société Royale de Médecine, 1786, cited in Corbin, *The Foul and the Fragrant*, p. 23.
7 Carl Bock, *Sundhedslære for folket* [Public Health Manual], Norwegian edn, tr. Henrik Bahr (B. C. Lønnings forlag, 1874), p. 103.
8 Gérard Vallée and Lynn McDonald, eds, *Florence Nightingale on Health in India* (Wilfrid Laurier University Press, 2006); Louise Penner, *Victorian Medicine and Social Reform: Florence Nightingale Among the Novelists* (Palgrave Macmillan, 2010), p. 17.
9 Haakon Natvig, *Lærebok i hygiene* [Hygiene Handbook] (Liv og helses forlag, 1958), p. 19.
10 Harold McGee, *Nose Dive: A Field Guide to the World's Smells* (Penguin Books, 2020), pp. 500–501.
11 Hallé, cited in Corbin, *The Foul and the Fragrant*, p. 29.
12 Corbin, *The Foul and the Fragrant*, p. 58.
13 Ibid., p. 27.
14 Anonymous poet in *Punch*, July–December 1847.
15 The French chemist, Michel Eugène Chevreul, cited in Corbin, *The Foul and the Fragrant*, p. 25.
16 Corbin, *The Foul and the Fragrant*, p. 30.
17 The French physician Philippe Passot, cited ibid., p. 26.
18 Natvig, *Lærebok i hygiene*, p. 21.
19 Erik Cohen, 'The Broken Cycle: Smell in a Bangkok Lane', in *The Smell Culture Reader*, ed. Jim Drobnick (Berg, 2006), pp. 118–27 (p. 118).
20 Victor Hugo, *Les Misérables* [1862], tr. Lee Fahnestock and Norman MacAfee, based on the original French to English translation by C. E. Wilbour (Signet Classics, 1987), p. 1257.

11 The Cast-Iron Plant: The Smell of Class

1 *Quarterly Review*, 30 (John Murray, 1823–4), p. 116.

2 A. de Capell Brooke, *Travels through Sweden, Norway, and Finmark to North Cape, in the Summer of 1820* (Rodwell and Martin, 1823), pp. 116–17.

3 Ibid., pp. 118–19.

4 Trygg Engen, *The Perception of Odors* (Academic Press, 1982).

5 Göran Stålbom, 'Ventilation och hälsa under 1900-talet' [Ventilation and Health in the 1900s], in *Människan inomhus. Perspektiv på vår tids inneliv. En antologi* [The Human Indoors: A Perspective on the Interior Life of Our Age], ed. Göran Stålbom and Birgitta Johansson (Formas, 2003), pp. 283–306 (p. 289).

6 Colin Trodd, 'Culture, Class, City: The National Gallery, London and the Spaces of Education, 1822–57', in *Art Apart: Art Institutions and Ideology Across England and North America*, ed. Marcia R. Pointon (Manchester University Press, 1994), pp. 42–3.

7 Mark Twain, 'About Smells', in *The Curious Republic of Gondour and Other Whimsical Sketches* [1871], Project Gutenberg, www.gutenberg.org, accessed 23 September 2025.

8 From Étienne Pivert de Senancour's novel, *Oberman*, cited in Alain Corbin, *The Foul and the Fragrant: Odor and the French Social Imagination* (Harvard University Press, 1986), p. 78.

9 Henrik Ibsen, *An Enemy of the People* [1882], tr. Robert Farquharson Sharp (Mockingbird Publishing, 2014).

10 Constance Classen, David Howes and Anthony Synnott, *Aroma: The Cultural History of Smell* (Routledge, 1994), p. 81.

11 Ibid.

12 Corbin, *The Foul and the Fragrant*, p. 71.

13 Henrikka Høegh Schmidt, *Veiledningsbok for tjenestepiker og unge husmødre* [Book of Guidelines for Serving Girls and Young Housewives] (Grøndahl & Søns forlag, 1912), pp. 11–12.

14 Classen, Howe and Synnott, *Aroma*, p. 168.

15 Chadwick, cited ibid., p. 83.

16 Bjørnstjerne Bjørnson, 'Holgar Drachmann', in *Poems and Songs of Bjørnstjerne Bjørnson*, tr. Arthur Hubbell Palmer (American-Scandinavian Foundation, 1915).

17 George Orwell, *Keep the Aspidistra Flying* [1936] (Penguin, 2014), p. 29.

18 David Howes, 'The Craft of the Senses', Concordia University Occasional Paper, 2011.
19 Corbin, *The Foul and the Fragrant*, p. 164.
20 George Orwell, *The Road to Wigan Pier* [1937] (Penguin, 1963), pp. 112–13.
21 Twain, 'About Smells'.
22 Ibid.

12 Damp Cows and Shallow Graves: The Smell of Confusion

1 Edgar Allan Poe, 'The Murders in the Rue Morgue', *Graham's Magazine*, 1841.
2 Arthur Conan Doyle, *The Hound of the Baskervilles* (George Newnes Ltd, 1902).
3 Melanie A. Kiechle, *Smell Detectives: An Olfactory History of Nineteenth-Century Urban America* (University of Washington Press, 2017), p. 36.
4 Dominique Laporte, *History of Shit* [1978], tr. Nadia Benabid and Rodolphe El-Khoury (MIT Press, 2002), p. xi.
5 Eir Grytli and Eli Støa, eds, *Fra årestue til smarthus – teknologien omformer boligen* [From Open-Hearth House to Smart-House – Technology Reshapes the Dwelling] (Norsk arkitekturforlag, 1998), p. 158.
6 Kiechle, *Smell Detectives*, p. 183.
7 Rodolphe El-Khoury, 'Polish and Deodorize: Paving the City in Late Eighteenth-Century France', in *The Smell Culture Reader*, ed. Jim Drobnick (Berg, 2006), pp. 18–28.
8 Eilert Sundt, *Om renligheds-stellet i Norge. Til oplysning om flid og fremskridt i landet* [On Cleanliness in Norway, 1869] (Bibliotek for de tusen hjem, 1887), p. 183.
9 Ibid., p. 12.
10 Haakon Natvig, *Lærebok i hygiene* [Hygiene Handbook] (Liv og helses forlag, 1958), p. 22.
11 Göran Stålbom, 'Ventilation och hälsa under 1900-talet' [Ventilation and Health in the 1900s], in *Människan inomhus. Perspektiv på vår tids inneliv. En antologi* [The Human Indoors: A Perspective on the Interior Life of Our Age], ed. Göran Stålbom and Birgitta Johansson (Formas, 2003), pp. 283–306 (p. 288).
12 Ibid., pp. 290–91.

13 Gustav Uddgren, *Strindberg the Man*, tr. Axel Johan Uppvall (The Four Seas Company, 1920), p. 77.
14 Åke Bresle, *Näsa för lukter* [A Nose for Smells] (Ingejörsförlag, 1986), p. 78.
15 Hans J. Rindisbacher, *The Smell of Books: A Cultural-Historical Study of Olfactory Perception in Literature* (University of Michigan Press, 1992).

13 Alabaster Brows: The Smell of Longing

1 Joris-Karl Huysmans, *Parisian Sketches* [1880], tr. Brendan King (Dedalus Ltd, 2014), p. 126.
2 Valerianate of ammonia was a medicine for the relief of nervous disorders; ibid., pp. 127–8.
3 Charles Baudelaire, *Les Fleurs du mal* [The Flowers of Evil, 1857], tr. James McGowan (Oxford University Press, 2008).
4 Per Buvik, *Dekadanse* [Decadence] (Pax forlag, 2001).
5 Joris-Karl Huysmans, *Against Nature* [1884], tr. Robert Baldick (Penguin Books, 1959).
6 Ibid., p. 36.
7 Ibid., p. 124.
8 Ibid.
9 Ibid., p. 220.
10 Robert de Montesquiou-Fézensac, *Pays des aromates* [Land of Scents] (H. Floury, Éditeur, 1900).
11 Edgar Munhall and James McNeill, *Whistler and Montesquiou: The Butterfly and the Bat* (Flammarion, 1995).
12 Thomas Mann, *Buddenbrooks* [1901], tr. H. T. Lowe-Porter (Vintage Books, 1996).
13 Danuta Fjellestad, 'Towards an Aesthetics of Smell, or, the Foul and the Fragrant in Contemporary Literature', *Cauce*, 24 (2001), pp. 637–51.
14 Hellen Giblin-Jowett, 'Smell, Smells and Smelling in Victorian Supernatural Fiction of the Fin de Siècle', PhD thesis, School of English Literature, Language and Linguistics, Newcastle University, 2014, p. 244.
15 Arne Garborg, *Weary Men* [1890], tr. Sverre Lyngstad (Northwestern University Press, 1999); Oscar Wilde, *The Picture of Dorian Gray* [1891] (Penguin Classics, 2010).

16 Marcel Proust, *In Search of Lost Time* [1913], tr. C. K. Scott Moncrieff and Terence Kilmartin, rev. D. J. Enright (The Modern Library, 2003).
17 James Joyce, *Ulysses* (Shakespeare, 1922).
18 Jan Bojer Vindheim, *Historiens hemmeligheter. Myte, magi og mysterier fra Atlantis til Harry Potter* [History's Secrets: Myth, Magic and Mystery from Atlantis to Harry Potter] (Kagge forlag, 2005).
19 Arthur Conan Doyle, *Memories and Adventures* (Hodder & Stoughton, 1924).

14 Interview with a Billy-Goat: The Smell of Denial

1 Maria Lind, 'What Is Wrong with My Nose: From Gogol and Freud to Goldin+Senneby (via Haraway)', *e-flux Journal*, 108 (April 2020), at www.e-flux.com.
2 Constance Classen, David Howes and Anthony Synnott, *Aroma: The Cultural History of Smell* (Routledge, 1994).
3 Sigmund Freud, *Civilization and Its Discontents*, tr. Joan Riviere (Hogarth Press, 1930).
4 Carol Mavor, '*Odor di Femina*: Though You May Not See Her, You Can Certainly Smell Her', in *The Smell Culture Reader*, ed. Jim Drobnick (Berg, 2006), pp. 277–88 (pp. 281–2).
5 Ibid., p. 283.
6 Drobnick, 'Preface[to Part v]', *The Smell Culture Reader*, p. 257.
7 Ibid.
8 Classen, Howes and Synnott, *Aroma*.
9 Hans J. Rindisbacher, *The Smell of Books: A Cultural-Historical Study of Olfactory Perception in Literature* (University of Michigan Press, 1992).
10 David Howes, 'The Social Life of the Senses', *Ars Vivendi Journal*, 3 (February 2013), pp. 4–23.
11 Victoria Henshaw, *Urban Smellscapes: Understanding and Designing City Smell Environments* (Routledge, 2014), p. 68.
12 Le Corbusier, *Towards a New Architecture*, tr. Frederick Etchells (Butterworth & Co., 1989); Le Corbusier, cited in Juhani Pallasmaa, *The Eyes of the Skin: Architecture and the Senses* (John Wiley & Sons Ltd, 2005), p. 22.
13 Friedrich Nietszche, cited in Beata Labuhn, 'Breathing a Moldy Air: Olfactory Experience, Aesthetics, and Ethics in the Writing of Ruskin and Riegl', *Future Anterior*, 13/2 (2016), pp. 103–17;

Jonathan Reinarz, *Past Scents: Historical Perspectives on Smell* (University of Illinois Press, 2014), p. 18.

14 Jim Drobnick and Jennifer Fisher, 'Perfumatives: Olfactory Dimensions in Contemporary Art', *Aroma-Chology Review*, 12/1 (1998), pp. 1, 4.

15 Caro Verbeek, *Ruiken aan de tijd. De olfactorische dimensie van het futurism (1909–1942)* [Smelling Time: The Olfactory Dimension of Futurism] (Vrije Universiteit Amsterdam, 2020).

16 Filippo Tommaso Marinetti, cited in Carol Verbeek, 'In Search of Lost Scents', in *International Yearbook of Futurism Studies*, vol. x, ed. Günter Berghaus (de Gruyter, 2021), pp. 247–75.

17 Cited in Larry Shiner, *Art Scents: Exploring the Aesthetics of Smell and the Olfactory Arts* (Oxford University Press, 2020).

18 Filippo Tommaso Marinetti, *The Futurist Cookbook* [1932], tr. Suzanne Brill (Penguin, 2015).

19 Ibid.

20 From Marinetti's manifesto for the Abyssinian War, cited in Walter Benjamin, 'The Work of Art in the Age of Mechanical Reproduction', in *Illuminations*, ed. Hannah Arendt, tr. Harry Zohn (Schocken Books, 1969), p. 241.

15 The Battlefield: The Smell of Insanity

1 Ernst Jünger, *Storm of Steel* [1920], tr. Michael Hofmann (Penguin Classics, 2004), p. 6.

2 *Apocalypse Now*, dir. Francis Ford Coppola (American Zoetrope, 1979).

3 Richard Humphreys, *Under Pressure: Living Life and Avoiding Death on a Nuclear Submarine* (HarperCollins, 2019).

4 Paul Amrozowicz, 'Submarine Life – What I Definitely Don't Miss', https://paulamrozowicz.com, 10 February 2015.

5 Peter Englund, *The Battle of Poltava: The Birth of the Russian Empire*, tr. Peter Hale (Victor Gollancz Ltd, 1992), p. 205.

6 Cited in Mark M. Smith, *The Smell of Battle, the Taste of Siege: A Sensory History of the Civil War* (Oxford University Press, 2015), p. 79.

7 Melanie A. Kiechle, *Smell Detectives: An Olfactory History of Nineteenth-Century Urban America* (University of Washington Press, 2017), p. 106.

8 Anthony J. Cumming, *The Battle for Britain: Interservice Rivalry between the Royal Air Force and the Royal Navy* (Naval Institute Press, 2015).

9 Lars Ericson Wolke, *Bombkrigets historia. Taktik och terror under 100 år av flygkrig* [The History of Bomb Warfare: Tactics and Terror Over a Century of Aerial Warfare] (Historiska Media, 2013).

10 Kurt Vonnegut, *Slaughterhouse-Five* (Dell, 1969).

11 Jünger, *Storm of Steel*, p. 93.

12 Organisation for the Prohibition of Chemical Weapons, www.opcw.org, accessed 17 September 2025.

13 Oskar Groening, cited in Jay Akbar, '"The brain and heart are attacked first . . . it begins with a stinging chest and epileptic seizures": Trial of the "Auschwitz Bookkeeper" Told How Gas Chamber Pesticide Was So Weak It Could Take Holocaust Victims "Longer than Half-an-Hour to Die"', *Daily Mail*, www.dailymail.co.uk, 27 May 2015.

14 R. L. Kirk and N. S. Stenhouse, 'Ability to Smell Solutions of Potassium Cyanide', *Nature*, 171/4355 (1953), pp. 698–9.

15 Pery Broad, cited in Hans J. Rindisbacher, *The Smell of Books: A Cultural-Historical Study of Olfactory Perception in Literature* (University of Michigan Press, 1992), p. 143.

16 Rindisbacher, *The Smell of Books*.

17 Johann Paul Kremer, cited ibid., p. 141.

18 Adolf Hitler, *Mein Kampf* [1925–6], tr. Ralph Manheim (Hutchinson, 1969).

19 Olga Lengyel, *Five Chimneys: A Woman Survivor's True Story of Auschwitz* (Academy Chicago Publishers, 1995), p. 160.

20 Devon E. Hinton et al., 'Olfactory-Triggered Panic Attacks Among Khmer Refugees', in *The Smell Culture Reader*, ed. Jim Drobnick (Berg, 2006), pp. 68–81 (p. 69).

16 Love Apples: The Smell of Lust

1 Amos Nadler et al., 'The Bull of Wall Street: Experimental Analysis of Testosterone and Asset Trading', *Management Science*, 64/9 (2017), pp. 4032–51.

2 No figure is available for the odour detection threshold in air, but based on the threshold for a solution of 20 ppb in water, we have

grounds to assume that it is low, albeit somewhat higher than for androstenone.

3 Erik Tunstad, 'Sexy lukt – eller frastøtende?' [Sexy Smell – or Disgusting?], www.forskning.no, 18 September 2007.
4 Preet Bano Singh et al., 'Smelling Anxiety Chemosignals Impairs Clinical Performance of Dental Students', *Chemical Senses*, 43/6 (2018), pp. 411–17.
5 Henry Miller, *Tropic of Cancer* [1934] (Grove Press, 1961).
6 Harold McGee, *Nose Dive: A Field Guide to the World's Smells* (Penguin Books, 2020), p. 116.
7 Eve Ensler, *The Vagina Monologues* (Villard, 2008).
8 W. D. Morrison, cited in Christopher Looby, 'The Roots of the Orchis, the Iuli of Chestnuts: The Odor of Male Solitude', in *The Smell Culture Reader*, ed. Jim Drobnick (Berg, 2006), pp. 289–304 (p. 290).
9 J. Kromer et al., 'Influence of HLA on Human Partnership and Sexual Satisfaction', *Scientific Reports*, 6/1 (2016), art. 32550, at www.nature.com.
10 McGee, *Nose Dive*, p. 117.
11 Kate Fox, *The Smell Report: An Overview of Facts and Findings* (Social Issues Research Centre, 1999).
12 Dan McKenzie, *Aromatics and the Soul: A Study of Smells* (William Heinemann, 1923), p. 84.
13 Jonathan Reinarz, *Past Scents: Historical Perspectives on Smell* (University of Illinois Press, 2014).
14 Alain Corbin, 'The New Calculus of Olfactory Pleasure', in *The Smell Culture Reader*, ed. Drobnick, pp. 167–80 (p. 178).
15 McKenzie, *Aromatics and the Soul*.
16 Diane Ackerman, *A Natural History of the Senses* (Random House, 1990), p. 24.
17 Ibid.
18 Paul Enzlin et al., '"To Shave or Not to Shave": Pubic Hair Removal and Its Association with Relational and Sexual Satisfaction in Women and Men', *Journal of Sexual Medicine*, 16/7 (2019), pp. 954–62.
19 Rebecca M. Herzig, *Plucked: A History of Hair Removal* (New York University Press, 2016).
20 Tony Perrottet, 'Secrets of the Great British Sex Clubs', *Slate*, 14 December 2009.

21 Lise Widding Isaksen, 'Kroppsutsondringers symbolsk makt' [The Secret Power of Bodily Secretions], *Tidsskrift for Den norske legeforening*, 1 (10 January 2002), pp. 79–81.
22 Günther Ohloff et al., *Scent and Chemistry: The Molecular World of Odors*, 2nd edn (Wiley-VCH, 2022).
23 Cited in Robert Jütte, 'Reodorizing the Modern Age', in *Smell and History: A Reader*, ed. Mark M. Smith (West Virginia University Press, 2019), pp. 170–86 (p. 177).
24 Slogan for Rexona deodorant in the 1960s and '70s.
25 PerfumersWorld, www.perfumersworld.com, accessed 17 September 2025.
26 Megan Williams and Coren Apicella, 'Synthetic Copulin Does Not Affect Men's Sexual Behavior', *Adaptive Human Behavior and Physiology*, 4 (2018), pp. 121–37.

17 White Sheets: The Smell of Purity

1 Serhij Zjadan, *Anarchy in the UKR* [2005], translated into Norwegian by Dagfinn Foldøy (Pax forlag, 2022).
2 J. Douglas Porteous, 'Smellscape', in *The Smell Culture Reader*, ed. Jim Drobnick (Berg, 2006), pp. 89–106 (p. 96).
3 Maria Pirogovskaya, 'Odour of Chlorine in Soviet Urban Reality: Notes on the Clean and the Collective', in *Les Cinq sens de la ville* [The Five Senses of the City], ed. Robert Beck, Ulrike Krampl and Emmanuelle Retaillaud-Bajac (Presses universitaires François-Rabelais, 2013), pp. 199–210.
4 Siv Ringdal, *Det amerikanske Lista. Med 110 volt i huset* [110 Volts at Home: The American Lista] (Pax forlag, 2002), p. 212.
5 Eva Ramm, *Med støv på hjernen* [Dust on the Brain] (Aschehoug, 1958), p. 79.
6 Michael S. Carolan, 'When Good Smells Go Bad: A Socio-Historical Understanding of Agricultural Odor Pollution,' *Environment and Planning A: Economy and Space*, 40/5 (2008), pp. 1235–49.
7 William Ian Miller, *The Anatomy of Disgust* (Harvard University Press, 1997).
8 Mikael Niemi, *Popular Music from Vittula*, tr. Laurie Thompson (Seven Stories Press, 2003), p. 211.

9 Ramm, *Med støv på hjernen*, p. 95.
10 Bjørn Berge, *The Ecology of Building Materials*, 2nd edn, tr. Chris Butters and Filip Henley (Taylor & Francis Group, 2009), p. 139.
11 Neeltje van Doremalen et al., 'Aerosol and Surface Stability of SARS-COV-2 as Compared with SARS-COV-1', *New England Journal of Medicine*, 382/16 (16 April 2020), pp. 1564–7.
12 Steve Ritter, 'New Car Smell: VOCs Account for the Characteristic "Newness"', *Chemical and Engineering News*, 80/20 (2002), p. 45.
13 Slogan for Ajax scouring powder in the 1960s.
14 Dominique Laporte, *History of Shit* [1978], tr. Nadia Benabid and Rodolphe El-Khoury (MIT Press, 2002), p. 83.
15 Anne Spurkland, *Immun – Kroppens evige kamp for å overleve* [Immune: The Body's Endless Struggle for Survival] (Spartacus, 2017).
16 Leena von Hertzen et al., 'Helsinki Alert of Biodiversity and Health', *Annals of Medicine*, 47/3 (2015), pp. 218–25.
17 A. E. Wold, 'The Hygiene Hypothesis Revised: Is the Rising Frequency of Allergy Due to Changes in the Intestinal Flora?', *Allergy*, 53 (1998), pp. 20–25.

18 The Blood of an Englishman: The Smell of the Unknown

1 Hsuan Hsu, *The Smell of Risk: Environmental Disparities and Olfactory Aesthetics* (New York University Press, 2020), p. 115.
2 Mark M. Smith, 'Making "Others" Smell', in *Smell and History: A Reader*, ed. Mark M. Smith (West Virginia University Press, 2019), pp. 187–202 (p. 188).
3 J. Douglas Porteous, 'Smellscape', in *The Smell Culture Reader*, ed. Jim Drobnick (Berg, 2006), pp. 89–106 (p. 93).
4 C. M. Woolgar, 'Medieval Smellscapes', in *Smell and History*, ed. Smith, pp. 50–75 (p. 62).
5 Dan McKenzie, *Aromatics and the Soul: A Study of Smells* (William Heinemann, 1923), p. 10.
6 Kate Fox, *The Smell Report: An Overview of Facts and Findings* (Social Issues Research Centre, 1999), p. 28.
7 Ibid.
8 Hsu, *The Smell of Risk*, p. 155.
9 Constance Classen, David Howes and Anthony Synnott, *Aroma: The Cultural History of Smell* (Routledge, 1994), p. 114.

10 Charles Baudelaire, 'Brussel Spleen', in the collection *Late Fragments; Flares, My Heart Laid Bare, Prose Poems, Belgium Disrobed*, tr. Richard Sieburth (Yale University Press, 2023), p. 227.
11 D. H. Lawrence, *Mornings in Mexico* (Martin Secker, 1927).
12 Jonathan Reinarz, *Past Scents: Historical Perspectives on Smell* (University of Illinois Press, 2014).
13 Fox, *The Smell Report*, pp. 32–3.
14 Jon Røyne Kyllingstad, *Rase. En vitenskapshistorie* [Race: A Scientific History] (Cappelen Damm, 2023), p. 110.
15 Constance Classen, 'The Odor of the Other: Olfactory Symbolism and Cultural Categories', *Ethos*, 20/2 (1992), pp. 133–66.
16 Martin F. Manalansan, 'Immigrant Lives and the Politics of Olfaction in the Global City', in *The Smell Culture Reader*, ed. Drobnick, pp. 41–52 (p. 46).
17 Ibid., p. 47.
18 Tullia Jack, 'Cleanliness and Consumption: Exploring Material and Social Structuring of Domestic Cleaning Practices', *International Journal of Consumer Studies*, 41/1 (2017), pp. 70–78.
19 Ibid., p. 76.
20 Alison Gerber, '"Everyone's Annoyed": Leveraging Uncertainty in the Smell of Others', *Cultural Sociology*, 16/3 (2022), p. 348.
21 Ibid.
22 Ibid., p. 350.
23 Aldous Huxley, *Those Barren Leaves* [1925] (Penguin, 1951), pp. 179–80.

19 Hygienic Equivalents: The Smell of Nothing

1 Bjørn Berge, *De siste syke hus* [The Last Sick Buildings] (Universitetsforlag, 1988).
2 Carl-Gustav Bornehag, 'Inomhusmiljö och hälsa – vad vet vi om sambanden?' [Indoor Environment and Health – What Do We Know About the Link?], in *Manniskan inomhus. Perspektiv pa var tids inneliv. En antologi* [The Human Indoors: A Perspective on the Interior Life of Our Age], ed. Göran Stalbom and Birgitta Johansson (Formas, 2003), pp. 133–54.
3 Anette Stenslund, 'A Whiff of Nothing: The Atmospheric Absence of Smell', *The Senses and Society*, 10/3 (2015), pp. 341–60; see also, for example, Axel Holst, *Hygiene* (H. Aschehoug & Co., 1928), pp. 35–6.

4 Povl Ole Fanger, 'Introduction of the Olf and the Decipol Units to Quantify Air Pollution Perceived by Humans Indoors and Outdoors', *Energy and Buildings*, 12/1 (1988), pp. 1–6.
5 Ibid.
6 Michael A. McGinley and Charles M. McGinley, 'The "Gray Line" Between Odor Nuisance and Health Effects', *Proceedings of the Air and Waste Management Association 92nd Annual Meeting and Exhibition* (St. Croix Sensory Inc., 1999).
7 Lisa Heschong, *Thermal Delight in Architecture* (MIT Press, 1979).
8 Edward T. Hall, *The Hidden Dimension* (Anchor Books, 1969).
9 Italo Calvino, *Six Memos for the Next Millennium*, tr. Patrick Creagh (Harvard University Press, 1988), p. 57.
10 Juhani Pallasmaa, *The Eyes of the Skin: Architecture and the Senses*, (John Wiley & Sons Ltd, 2005).
11 Nationwide survey conducted by Respons Analyse on behalf of the GK engineering firm, June 2022.
12 Carey J. Simonson, *Moisture, Thermal and Ventilation Performance of Tapanila Ecological House* (VTT, 2000).

20 The Scent Organ: The Smell that Remains

1 Aldous Huxley, *Brave New World* [1932] (Chatto & Windus, 1947), p. 127.
2 Ibid., p. 41.
3 Ibid., p. 196.
4 Günther Ohloff et al., *Scent and Chemistry: The Molecular World of Odors*, 2nd edn (Wiley-VCH, 2022), p. 106.
5 Mark W. D. Paterson, 'Digital Scratch and Virtual Sniff: Simulating Scents', in *The Smell Culture Reader*, ed. Jim Drobnick (Berg, 2006), pp. 358–67.
6 Elise Kjørstad, 'Slik får lukt og lyd oss til å kjøpe mer' [How Smells and Sounds Make Us Buy More], www.forskning.no, 28 November 2017.
7 David Hendy, *Noise: A Human History of Sound and Listening* (Profile Books, 2013), p. 265.
8 Jacob Leander Olsson and Ann Wenehed, 'The Smell of Money: A Study of How Scents Can Affect Behavior and Attitudes', Master's thesis, Stockholm School of Economics, 2010.

9 Jim Drobnick, 'Eating Nothing: Cooking Aromas in Art and Culture', in *The Smell Culture Reader*, ed. Drobnick, pp. 342–56 (p. 351).

10 Hsuan Hsu, *The Smell of Risk: Environmental Disparities and Olfactory Aesthetics* (New York University Press, 2020), p. 51.

11 Till Luckenbach and David Epel, 'Nitromusk and Polycyclic Musk Compounds as Long-Term Inhibitors of Cellular Xenobiotic Defense Systems Mediated by Multidrug Transporters', *Environmental Health Perspectives*, 113/1 (January 2005), pp. 17–24.

12 Anne Lise Stranden, 'Lukten i butikken kan berolige kjøperne' [The Smell in Shops Can Calm Shoppers], www.forskning.no, 24 June 2014.

13 Peter Damian and Kate Damian, 'Environmental Fragrancing', in *The Smell Culture Reader*, ed. Drobnick, pp. 148–60 (pp. 152, 155); Jonathan Haidt, 'The Emotional Dog and Its Rational Tail: A Social Intuitionist Approach to Moral Judgment', *Psychological Review*, 108/4 (2001), pp. 814–34.

14 Constance Classen, David Howes and Anthony Synnott, *Aroma: The Cultural History of Smell* (Routledge, 1994), p. 205.

15 José Saramago, *Blindness*, tr. Giovanni Pontiero (Harcourt Brace & Company, 1998), p. 126.

Afterword

1 Per Christian Høydalsvik, 'Den immaterielle erfaringen. En undersøkelse av utstillingen Sissel Tolaas RE____ 08.10.21–30.12.21 ved Astrup Fearnley Museet' [The Immaterial Experience: A Study of Sissel Tolaas's Exhibition RE____ 08.10.21–30.12.21 at the Astrup Fearnley Museum], Master's thesis, OsloMet, 2022, www.oslomet.no, accessed 19 September 2025.

2 Elmar Schmeisser, Kimberly A. Pollard and Tomasz Letowski, *Olfaction Warfare: Odor as Sword and Shield* (Army Research Laboratory, 2013).

3 Jung Chang, *Empress Dowager Cixi* (Alfred A. Knopf, 2013), p. 272.

BIBLIOGRAPHY

Abdulsalam, Oluwatosin, et al. (2021). 'Phytohormones and Volatile Organic Compounds, Like Geosmin, in the Ectomycorrhiza of *Tricholoma vaccinum* and Norway Spruce (*Picea abies*)'. Article in *Mycorrhiza*, 31, pp. 173–88.

Ackerman, Diane (1990). *A Natural History of the Senses*. New York: Random House.

Adams, Sean (2021). 'John Cleese on Dark Comedy, His Best Monty Python Moment, and His Upcoming Show at Hershey'. Published at www.pennlive.com, 7 October.

Akbar, Jay (2015). '"The brain and heart are attacked first . . . it begins with a stinging chest and epileptic seizures": Trial of "Auschwitz Bookkeeper" Told How Gas Chamber Pesticide Was So Weak It Could Take Holocaust Victims "Longer Than Half-an-Hour To Die"'. Article in the *Daily Mail*, www.dailymail.co.uk, 27 May.

Alho, Laura, et al. (2015). 'Nosewitness Identification: Effects of Negative Emotion'. Article in *PLOS One*, 10/1 (22 January), at https://journals.plos.org.

Amrozowicz, Paul (2015). 'Submarine Life – What I Definitely Don't Miss', https://paulamrozowicz.com, 10 February.

Anonymous (1847). 'The Vampyre'. Published in *Punch*, July–December.

Arshamian, A., et al. (2022). 'The Perception of Odor Pleasantness Is Shared Across Cultures'. Article in *Current Biology*, 32/9, pp. 2061–6.

Asphaug, Helene (2022). 'Stivpynta feminister. Du er ikke frigjort når du bruker mer tid på å bli sett enn hørt' [You Are Not Liberated if You Spend More Time on Being Seen than Heard]. *NRK Ytring*, 23 April.

Auguet, Roland (1994). *Cruelty and Civilization: The Roman Games*. London: Routledge.

Bachelard, Gaston (1969 [1964]). *The Poetics of Space*. Translated by Maria Jolas. Boston: Beacon Press.

Bandlien, Bjørn (2011). 'Bever med baller' [Beavers with Balls]. Article in *Norsk Folkeminnelag*, 58, pp. 20–23.

Bang-Andersen, Sveinung (1983). *Svarthåla på Viste – boplass i 6000 år* [The Black Cave – A Stone Age Dwelling Place]. Stavanger: AmS-Småtrykk.

Barwich, A. S. (2020). *Smellosophy: What the Nose Tells the Mind*. Cambridge, MA: Harvard University Press.

Baudelaire, Charles (2008 [1857]). *Les Fleurs du mal* [The Flowers of Evil]. Translated by James McGowan. Oxford: Oxford University Press.

— (2023). *Late Fragments: Flares, My Heart Laid Bare, Prose Poems, Belgium Disrobed*. Translated into English by Richard Sieburth. New Haven: Yale University Press.

Benjamin, Walter (1969). 'The Work of Art in the Age of Mechanical Reproduction'. Essay in *Illuminations*, ed. Hannah Arendt. New York: Schocken Books, pp. 217–51.

Bentzrød, Sveinung Berg (2021). 'De norske Kabul-historiene' [The Norwegian Kabul Stories]. Article in *Aftenposten*, www.aftenposten.no, 3 September.

Berge, Bjørn (1988). *De siste syke hus* [The Last Sick Buildings]. Oslo: Universitetsforlaget.

— (2009). *The Ecology of Building Materials*. Second Edition. London: Taylor & Francis Group.

Bingen, Hildegard von (1997). *The Book of the Rewards of Life*. Translated by Bruce W. Hozeski. Oxford: Oxford University Press.

Bjørkvik, Harald (2017). *Byar og bysamfunn. Om hvordan byar er blitt til* [Cities and City Societies: On How Cities Came into Being]. Lillesandmuseet.

Bjørnson, Bjørnstjerne (1915). *Poems and Songs of Bjørnstjerne Bjørnson*. Translated by Arthur Hubbell Palmer. New York: American-Scandinavian Foundation.

Bock, Carl (1874). *Sundhedslære for folket* [Public Health Manual]. Translated into Norwegian by Henrik Bahr. Stavanger: B. C. Lønnings forlag.

Bradley, Mark, ed. (2015). *Smell and the Ancient Senses*. London: Routledge.

Bresle, Åke (1986). *Näsa för lukter* [A Nose for Smells]. Falköping: Ingenjörsförlaget.

Buvik, Per (2001). *Dekadanse* [Decadence]. Oslo: Pax forlag.
Byrne, Joseph P. (2006). *Daily Life During the Black Death*. Westport: Greenwood Press.
Calvino, Italo (1988). *Six Memos for the Next Millennium*. Translated by Patrick Creagh. Cambridge, MA: Harvard University Press.
Campbell, Gordon (2013). *The Hermit in the Garden: From Imperial Rome to Ornamental Gnome*. Oxford: Oxford University Press.
Capell Brooke, A. de (1823). *Travels through Sweden, Norway and Finmark to North Cape in the Summer of 1820*. London: Rodwell and Martin.
Carolan, Michael S. (2006). 'When Good Smells Go Bad: A Socio-historical Understanding of Agricultural Odor Pollution'. Article in *Environment and Planning A*, 40/5, pp. 1235–49.
Cervantes Saavedra, Miguel de (2009). *Don Quixote*. Translated by James H. Montgomery. Indianapolis/Cambridge, MA: Hackett Publishing Company.
Chang, Jung (2013). *Empress Dowager Cixi*. New York: Alfred A. Knopf.
Chrisafis, Angelique (2015). 'Proust's Memory-Laden Madeleine Cakes Started Life as Toast, Manuscripts Reveal'. Article in *The Guardian*, www.theguardian.com, 19 October.
Christensen, Arne Lie (1995). *Den norske byggeskikken. Hus og bolig på landsbygda fra middelalderen til vår egen tid* [Norwegian Building Traditions. Houses and Homes in Villages from the Middle Ages to Our Own Times]. Oslo: Pax forlag A/S.
Clark, R.E.D. (1953). 'The Spheres of Revelation and Science: What Are Their Limitations in Relation to Each Other? Article in *Journal of the American Scientific Affiliation*, 5 June, pp. 8–17.
Classen, Constance (1992). 'The Odor of the Other: Olfactory Symbolism and Cultural Categories'. Article in *Ethos*, 20/2 (June), pp. 133–66.
— (1998). *The Color of Angels: Cosmology, Gender and the Aesthetic Imagination*. London: Routledge.
—, David Howes and Anthony Synnott (1994). *Aroma: The Cultural History of Smell*. London: Routledge.
Cobb, Matthew (2020). *Smell: A Very Short Introduction*. Oxford: Oxford University Press.
Conan Doyle, Arthur (1902). *The Hound of the Baskervilles*. London: George Newnes Ltd.
— (1924). *Memories and Adventures*. London: Hodder & Stoughton.

Corbin, Alain (1986). *The Foul and the Fragrant: Odor and the French Social Imagination*. Cambridge, MA: Harvard University Press.

Cumming, Anthony J. (2015). *The Battle for Britain: Interservice Rivalry between the Royal Air Force and the Royal Navy*. Annapolis: Naval Institute Press.

Dante Alighieri (1867). *The Divine Comedy. Inferno*. Translated by Henry Wadsworth Longfellow. Boston and New York: Houghton, Mifflin and Company.

Defoe, Daniel (1896). *A Journal of the Plague Year*. New York: Longmans, Green and Co.

Dickens, Charles (1905). *The Uncommercial Traveller*. London: Chapman & Hall.

Donovan, Michael (1830). *Domestic Economy*, vol. 1: *Brewing, Distilling, Wine-Making, Baking. &c.* London: Longman, Rees, Orme, Brown and Green.

Doremalen, Neeltje van, et al. (2020). 'Aerosol and Surface Stability of SARS-COV-2 as Compared with SARS- COV -1'. *New England Journal of Medicine*, 382/16 (April), pp. 1564–7.

Dostoevsky, Fyodor. (2007). *The Karamazov Brothers*. Translated by Constance Garnett. Ware, Hertfordshire: Wordsworth Editions.

Drobnick, Jim, ed. (2006). *The Smell Culture Reader*. Oxford: Berg.

Drobnick, Jim, and Jennifer Fisher (1998). 'Perfumatives: Olfactory Dimensions in Contemporary Art'. Article in *Aroma-Chology Review*, 12/1, pp. 1, 4.

Ehrenberg, Rachel (2015). 'Urban Microbes Come Out of the Shadows'. Article in *Nature*, 522/7557, pp. 399–400.

Elias, Norbert (1939). *The Civilizing Process: The History of Manners and State Formation and Civilization*. Oxford: Blackwell.

Emmerson, Allison L. C. (2020). *Life and Death in the Roman Suburb*. Oxford: Oxford University Press.

Engen, Trygg (1979). 'The Origin of Preferences in Taste and Smell'. Essay in *Preference Behaviour and Chemoreception*, ed. J.H.A. Kroeze. London: Information Retrieval, pp. 263–73.

— (1982). *The Perception of Odors*. New York: Academic Press.

Englund, Peter (1992). *The Battle of Poltava*. Translated by Peter Hale. London: Victor Gollancz Ltd.

— (1992). *Fortidens landskap. Om mordet på Karl XII og andre essays* [The Landscapes of the Past: On the Murder of Charles XII and

Other Essays]. Translated into Norwegian by Per Qvale. Oslo: Universitetsforlaget.

Ensler, Eve (2008). *The Vagina Monologues*. New York: Villard.

Enzlin, Paul Kaat Bollen, et al. (2019). '"To Shave or Not to Shave": Pubic Hair Removal and Its Association with Relational and Sexual Satisfaction in Women and Men'. Article in *Journal of Sexual Medicine*, 16/7 (July), pp. 954–62.

Fanger, Pavl Ole (1988). 'Introduction of the Olf and the Decipol Units to Quantify Air Pollution Perceived by Humans Indoors and Outdoors'. Article in *Energy and Buildings*, 12/1, pp. 1–6.

Fjellestad, Danuta (2001). 'Towards an Aesthetics of Smell, or, the Foul and the Fragrant in Contemporary Literature'. Article in *Cauce*, 24, pp. 637–51.

Foucault, Michel (1978). *The History of Sexuality*, vol. 1*: An Introduction*. Translated by Robert Hurley. New York: Pantheon Books.

Fox, Kate (1999). *The Smell Report: An Overview of Facts and Findings*. Oxford: Social Issues Research Centre (SIRC).

Francis, Gemma (2015). 'Freshly Baked Bread Tops Poll of Britain's Top 50 Favourite Smells'. Article in the *Daily Mirror*, www.mirror.co.uk, 25 May.

Fredriksen, Benjamin (2015). 'Jobben til Ansgar er å lukte på tørrfisk: – Det lukter gull' [Ansgar's Job Is to Sniff Dried Fish: It Smells of Gold]. Internet article, *NRK Nordland*, www.nrk.no, 30 August.

Freud, Sigmund (1930). *Civilization and Its Discontents*. Translated by Joan Riviere. London: Hogarth Press.

Frumin, Idan, et al. (2015). 'A Social Chemosignaling Function for Human Handshaking'. *eLife*, 4, art. e05154.

Fuchs, Thomas (2000). *A Concise Biography of Adolf Hitler*. New York: The Berkley Publishing Group.

Gade, Kari Ellen (1995). 'Einarr Þambarskelfir's Last Shot'. Article in *Scandinavian Studies*, 67/2, pp. 153–62.

Garborg, Arne (1999). *Weary Men*. Translated by Sverre Lyngstad. Evanston: Northwestern University Press.

Gerber, Allison (2022). '"Everyone's Annoyed": Leveraging Uncertainty in the Smell of Others'. Article in *Cultural Sociology*, 16/3, pp. 338–57.

Giblin-Jowett, Hellen (2014). 'Smell, Smells and Smelling in Victorian Supernatural Fiction of the Fin de Siècle'. Phd thesis, School of English Literature, Language and Linguistics. Newcastle University.

Gilbert, N. Avery, and Charles J. Wysocki (1987). 'The Smell Survey Results'. Article in *National Geographic*, 172 (October), pp. 514–25.

Gogol, Nikolai (2015). *The Nose*. Translated by Ronald Wilks. London: Penguin.

Grytli, Eir, and Eli Støa, eds (1998). *Fra årestue til smarthus – teknologien omformer boligen* [From Open-Hearth House to Smart-House – Technology Reshapes the Dwelling]. Oslo: Norsk arkitekturforlag.

Hagen, Rune Blix (2010). *Hekser. Fra forfølgelse til fortryllelse* [Witches: From Persecution to Enchantment]. Oslo: Humanist forlag.

Haidt, Jonathan (2001). 'The Emotional Dog and Its Rational Tail: A Social Intuitionist Approach to Moral Judgment'. Article in *Psychological Review*, 108/4, pp. 814–34.

Hall, Edward T. (1990). *The Hidden Dimension*. New York: Anchor Books Editions.

Hendy, David (2013). *Noise: A Human History of Sound and Listening*. London: Profile Books Ltd.

Henshaw, Victoria (2014). *Urban Smellscapes: Understanding and Designing City Smell Environments*. New York: Routledge.

Hertzen, Leena von, et al. (2015). 'Helsinki Alert of Biodiversity and Health'. Article in *Annals of Medicine*, 47/3 (2015), pp. 218–25.

Herz, Rachel (2007). *The Scent of Desire: Discovering Our Enigmatic Sense of Smell*. New York: William Morrow.

— (2012). *That's Disgusting: Unraveling the Mysteries of Repulsion*. New York: W. W. Norton.

Herzig, Rebecca M. (2016). *Plucked: A History of Hair Removal*. New York: New York University Press.

Heschong, Lisa (1979). *Thermal Delight in Architecture*. Cambridge, MA: MIT Press.

Hitler, Adolf (1969 [1925–6]). *Mein Kampf*. Translated by Ralph Manheim. London: Hutchinson.

Holst, Axel (1928). *Hygiene*. Oslo: H. Aschehoug & Co.

Howes, David, ed. (2004). *Empire of the Senses: The Sensual Culture Reader*. London: Routledge.

— (2011). 'The Craft of the Senses'. Occasional Paper, Concordia University.

— (2013). 'The Social Life of the Senses'. Article in *Ars Vivendi Journal*, 3 (February), pp. 4–23.

Hsu, Hsuan L. (2020). *The Smell of Risk: Environmental Disparities and Olfactory Aesthetics*. New York: New York University Press.

Hugo, Victor (1987). *Les Misérables*. Translated by C. E. Wilbour, Lee Fahnestock and Norman MacAfee. New York: Signet Classics.

Humphreys, Richard (2019). *Under Pressure: Living Life and Avoiding Death on a Nuclear Submarine*. London: HarperCollins.

Huxley, Aldous (1925). *Those Barren Leaves*. London: Chatto & Windus.

— (1947 [1932]). *Brave New World*. London: Chatto & Windus.

Huysmans, Joris-Karl (1959 [1884]). *Against Nature*. Translated by Robert Baldick. London: Penguin Books.

— (2014). *Parisian Sketches*. Translated by Brendan King. Sawtry: Dedalus Ltd.

Høydalsvik, Per Christian (2022). 'Den immaterielle erfaringen. En undersøkelse av utstillingen Sissel Tolaas RE____08.10.21 - 30.12.21 ved Astrup Fearnley Museet' [The Immaterial Experience: A Study of Sissel Tolaas's Exhibition RE____ 08.10.21–30.12.21 at the Astrup Fearnley Museum]. Master's thesis, OsloMet, www.oslomet.no.

Ibsen, Henrik (2014 [1882]). *An Enemy of the People*. Translated by Robert Farqharson Sharp. Middleton: Mockingbird Publishing.

Igarashi, Kei M., et al. (2014). 'Coordination of Entorhinal–Hippocampal Ensemble Activity During Associative Learning'. Article in *Nature*, 510, pp. 143–7.

Isaksen, Lise Widding (2002). 'Kroppsutsondringers symbolske makt' [The Secret Power of Bodily Secretions]. Article in *Tidsskrift for Den norske Legeforening*, 1 (10 January), pp. 79–81.

Jack, Tullia (2017). 'Cleanliness and Consumption: Exploring Material and Social Structuring of Domestic Cleaning Practices'. Article in *International Journal of Consumer Studies*, 41/1 (January), pp. 70–78.

Jansen, Gemma (2018). 'Sewers or Cesspits? About Modern Assumptions and Roman Preferences'. Essay in *Latrinae, Roman Toilets in the Northwestern Provinces of the Roman Empire*, ed. S. Hoss. Roman Archaeology Series. Oxford: Archaeopress, pp. 5–18.

Jordal, Vigdis, Anette Skuggedal and Ellen K. Syverstad, eds (2008). *Luft og kjærlighet* [Air and Love]. Oslo: Dinamo forlag.

Joyce, James (1992 [1916]). *Portrait of the Artist as a Young Man*. Ware, Hertfordshire: Wordsworth Editions.

— (1922). *Ulysses*. Paris: Shakespeare.

Jünger, Ernst (2004 [1920]). *Storm of Steel*. Translated by Michael Hofmann. London: Penguin Classics.

Jørgensen, Dolly (2014). 'Modernity and Medieval Muck'. Article in *Nature and Culture*, 9/3, pp. 225–37.

Kant, Immanuel (1974 [1798]). *Anthropology from a Pragmatic Point of View*. Translated by Mary J. Gregor. The Hague: Martinus Nijhoff.

Kazantzakis, Nikos (1965 [1957]). *Report to Greco*. Translated by P. A. Bien. New York: Simon & Schuster.

Kiechle, Melanie A. (2017). *Smell Detectives: An Olfactory History of Nineteenth-Century Urban America*. Seattle: University of Washington Press.

King, Anya H. (2017). *Scent from the Garden of Paradise: Musk and the Medieval Islamic World*. Islamic History and Civilization series. Leiden: Brill.

Kirk, R., and Stenhouse, N. (1953). 'Ability to Smell Solutions of Potassium Cyanide'. Article in *Nature*, 171/4355, pp. 698–9.

Kjørstad, Elise (2017). 'Slik får lukt og lyd oss til å kjøpe mer' [How Smells and Sounds Make Us Buy More]. Internet article, www.forskning.no, 28 November.

Klima- og forurensningsdirektoratet (2013). *Regulering av luktutslipp i tillatelser etter forurensningsloven* [Regulation of Odour Emissions in Permits as per the Pollution Control Act]. Veileder TA-3019/2013. Oslo.

Knausgård, Karl Ove (2009). *A Death in the Family: My Struggle Book 1*. Translated by Don Bartlett. London: Harvill Secker.

Kromer, J., et al. (2016). 'Influence of HLA on Human Partnership and Sexual Satisfaction'. Article in *Scientific Reports*, 6/1, art. 32550, at www.nature.com.

Kwak, Jae, et al. (2013). 'Changes in Volatile Compounds of Human Urine as It Ages: Their Interaction with Water'. Article in *Journal of Chromatography B*, 941, pp. 50–53.

Kyllingstad, Jon Røyne (2023). *Rase. En vitenskapshistorie* [Race: A Scientific History]. Oslo: Cappelen Damm.

Labuhn, Beata (2016). 'Breathing a Moldy Air: Olfactory Experience, Aesthetics, and Ethics in the Writing of Ruskin and Riegl'. Article in *Future Anterior*, 13/2, pp. 103–17.

Lakhani, Ali, et al. (2019). 'Using the Natural Environment to Address the Psychosocial Impact of Neurological Disability: A Systematic Review'. Article in *Health and Place*, 55 (January), pp. 188–201.

Lanciani, Rodolfo (1898), *Ancient Rome in the Light of Recent Discoveries*. Twelfth Edition. Boston: Houghton Mifflin & Company.

Laporte, Dominique (2002 [1978]). *History of Shit.* Translated by Nadia Benabid and Rodolphe el-Khoury. Cambridge, MA: MIT Press.

Last, John (2022). 'The Centuries-Long Quest for The Scent of God'. Internet article in *Noēma*, www.noemamag.com, 28 July.

Lawrence, D. H. (1927). *Mornings in Mexico*. London: Martin Secker.

Laxness, Halldór Kiljan (2008 [1934–5]). *Independent People.* Translated by J. A. Thompson. London: Vintage Books.

Le Corbusier (1989). *Towards a New Architecture*. Translated by Frederick Etchells. London: Butterworth & Co.

Lengyel, Olga (1995). *Five Chimneys: A Woman Survivor's True Story of Auschwitz.* Chicago: Academy Chicago Publishers.

Levin, David Michael, ed. (1993). *Modernity and the Hegemony of Vision*. Berkeley: University of California Press.

Lind, Maria (2020). 'What Is Wrong with My Nose: From Gogol and Freud to Goldin+Senneby (via Haraway)'. Article in *e-flux Journal*, 108 (April), at www.e-flux.com.

Lyotard, Jean-François (1988). *L'Inhumain: Causeries sur le temps.* Paris: Éditions Galilée.

Luckenbach, Till, and David Epel (2005). 'Nitromusk and Polycyclic Musk Compounds as Long-Term Inhibitors of Cellular Xenobiotic Defense Systems Mediated by Multidrug Transport'. Article in *Environmental Health Perspectives*, 113/1 (January), pp. 17–24.

McGee, Harold (2020). *Nose Dive: A Field Guide to the World's Smells.* New York: Penguin Press.

McGinley, Michael A., and Charles M. McGinley (1999). 'The "Gray Line" Between Odor Nuisance and Health Effects', *Proceedings of the Air and Waste Management Association 92nd Annual Meeting and Exhibition*. Stillwater: St. Croix Sensory Inc.

McKenzie, Dan (1923). *Aromatics and the Soul: A Study of Smells.* London: William Heinemann.

Mackenzie, George Steuart (1811). *Travels in the Island of Iceland, During the Summer of the Year MDCCCX*. Edinburgh: Archibald Constable and Company.

Mann, Thomas (1996 [1901]). *Buddenbrooks*. Translated by H. T. Lowe-Porter. London: Vintage Books.

Marinetti, Filippo Tommaso (2015 [1932]). *The Futurist Cookbook.* Translated by Suzanne Brill. London: Penguin.

Martin, R. M. (1832). *The Past and Present State of the Tea Trade of England, and of the Continents of Europe and America.* London: Parbury, Allen & Co.

Miller, Henry (1961 [1934]). *Tropic of Cancer*. New York: Grove Press.

Miller, Hugh (1869). *First Impressions of England and Its People.* Edinburgh: William P. Nimmo.

Miller, William Ian (1997). *The Anatomy of Disgust*. Cambridge, MA: Harvard University Press.

Montesquiou-Fézensac, Robert de (1900). *Pays des aromates* [Land of Scents]. Paris: H. Floury, Éditeur.

Moran, James K., et al. (2015). 'The Scent of Blood: A Driver of Human Behavior?' Article in *PLOS One*, 10/9 (23 September 2015), at https://journals.plos.org.

Muchembled, Robert (2020). *Smells: A Cultural History of Odours in Early Modern Times.* Translated by Susan Pickford. Cambridge: Polity Press.

Munhall, Edgar, and James McNeill (1995). *Whistler and Montesquiou: The Butterfly and the Bat.* Paris: Flammarion.

Nadler, Amos, et al. (2017). 'The Bull of Wall Street: Experimental Analysis of Testosterone and Asset Trading'. Article in *Management Science*, 64/9, pp. 4032–51.

Natvig, Haakon (1958). *Lærebok i hygiene* [Hygiene Handbook]. Oslo: Liv og Helses forlag (Institusjonen Scheibler).

Newcastle upon Tyne Town Council (1839). *Proceedings and Reports of the Town Council of the Borough of Newcastle for 1839.*

Niemi, Mikael (2003). *Popular Music from Vittula.* Translated by Laurie Thompson. New York: Seven Stories Press.

Nohain, Jean, and Françoise Caradec (1967). *Le Pétomane: 1857–1945.* Translated by Warren Tute. Los Angeles: Sherbourne Press.

Odden, Per Einar (2001). *Den hellige Kristina den Forbløffende (av Belgia) (1150–1224)* [St Christina the Astonishing of Belgium]. Published on the website of the Catholic Church of Norway, www.katolsk.no, 16 February.

Ohloff, Günther, et al. (2022). *Scent and Chemistry: The Molecular World of Odors.* Second Edition. Weinheim: Wiley-VCH.

Olsson, Jacob Leander, and Ann Wenehed (2010). 'The Smell of Money: A Study of How Scents Can Affect Behavior and Attitudes'. Master's thesis in Marketing, Stockholm School of Economics, Spring 2010.

Orwell, George (2014 [1936]). *Keep the Aspidistra Flying*. London: Penguin.
— (1963 [1937]). *The Road to Wigan Pier*. London: Penguin.
Pallasmaa, Juhani (2005). *The Eyes of the Skin: Architecture and the Senses*. Chichester: John Wiley & Sons Ltd.
Penner, Louise (2010). *Victorian Medicine and Social Reform: Florence Nightingale Among the Novelists*. New York: Palgrave Macmillan.
Perrottet, Tony (2009). 'Secrets of the Great British Sex Clubs'. Internet article in *Slate*, https://slate.com, 14 December.
Pirogovskaya, Maria (2013). 'Odour of Chlorine in Soviet Urban Reality: Notes on the Clean and the Collective'. Essay in *Les cinq sens de la ville* [The Five Senses of the City], ed. Robert Beck, Ulrike Krampl and Emmanuelle Retaillaud-Bajac. Tours: Presses Universitaires Francois-Rabelais, pp. 199–210.
Poe, Edgar Allan (1841). 'The Murders in the Rue Morgue'. *Graham's Magazine*.
Proust, Marcel (2003 [1913]). *In Search of Lost Time*. Translated by C. K. Scott Moncrieff and Terence Kilmartin, revised by D. J. Enright. New York: The Modern Library.
Qing-hong, Ji (2020). 'Marco Polo and Musk: On the Authenticity of Marco Polo's Visit to China'. Article in *China Tibetology*, 1, pp. 60–67.
Quarterly Review (1823–4). 'Art. v. – Travels through Sweden, Norway, and Finmark to the North Cape, in the Summer of 1820. By A. de Capell Brooke, M. A. London. 1823'. *Quarterly Review*, 30 (October 1823 and January 1824). London: John Murray, pp. 115–33.
Ramm, Eva (1958). *Med støv på hjernen* [Dust on the Brain]. Oslo: H. Aschehoug & Co.
Reinarz, Jonathan (2014). *Past Scents: Historical Perspectives on Smell*. Urbana: University of Illinois Press.
Riis, Annie (1994). *Satirens tid. Dikt og gjendiktninger* [The Age of Satire: Poems and Translations]. Oslo: Aschehoug & Co.
Rijneveld, Marieke Lucas (2020). *The Discomfort of Evening*. Translated by Michele Hutchinson. London: Faber & Faber.
Rilke, Rainer Marie (1989 [1910]). *The Notebooks of Malte Laurids Brigge*. Translated by Stephen Mitchell. New York: Vintage Books.
Rindisbacher, Hans J. (1992). *The Smell of Books: A Cultural-Historical Study of Olfactory Perception in Literature*. Ann Arbor: The University of Michigan Press.

Ringdal, Siv (2002). *Det amerikanske Lista. Med 110 volt i Huset* [110 Volts at Home: The American Lista]. Oslo: Pax forlag.
Ritter, Steve (2002). 'New Car Smell: VOCs Account for the Characteristic "Newness"'. Article in *Chemical and Engineering News*, 80/20, p. 45.
Rousseau, Jean-Jacques (2012 [1762]). *Of the Social Contract and Other Political Writings*. Translated by Quintin Hoare. London: Penguin.
Saramago, José (1998 [1995]). *Blindness*. Translated by Giovanni Pontiero. San Diego, New York, London: Harcourt Brace & Company.
Schmeisser, Elmar, Kimberly A. Pollard and Tomasz Letowski (March 2013). *Olfaction Warfare: Odor as Sword and Shield*. Report ARL-SR-0258. U.S. Army Research Laboratory.
Schmidt, Henrikka Høegh (1912). *Veiledningsbok for Tjenestepiker og unge Husmødre* [Book of Guidelines for Serving Girls and Young Housewives]. Kristiania: Grøndahl & Søns forlag.
Schønberg Erken, Henriette (1941 [1914]). *Stor kokebok for større og mindre husholdninger* [Big Cookbook for Households Large and Small]. Oslo: Aschehoug & Co.
Shiner, Larry (2020). *Art Scents: Exploring the Aesthetics of Smell and the Olfactory Arts*. Oxford: Oxford University Press.
Simonson, Carey J. (2000). *Moisture, Thermal and Ventilation Performance of Tapanila Ecological House*. Research Notes. Espoo: VTT Technical Research Centre of Finland.
Sing, Preet Bano, et al. (2018). 'Smelling Anxiety Chemosignals Impairs Clinical Performance of Dental Students'. Article in *Chemical Senses*, 43/6 (July), pp. 411–17.
Smith, Mark M. (2015). *The Smell of Battle, the Taste of Siege: A Sensory History of the Civil War*. Oxford: Oxford University Press.
— ed. (2019). *Smell and History: A Reader*. Morgantown: West Virginia University Press.
Spurkland, Anne (2017). *Immun – Kroppens evige kamp for å overleve* [Immune: The Body's Endless Struggle for Survival]. Oslo: Spartacus.
Stenslund, Anette (2015). 'A Whiff of Nothing: The Atmospheric Absence of Smell'. Article in *The Senses and Society*, 10/3, pp. 341–60.
Sterer, Nir, and Mel Rosenberg (2011). *Breath Odors: Origin, Diagnosis, and Management*. Heidelberg: Springer-Verlag.
Stranden, Anne Lise (2014). 'Lukten i butikken kan berolige kjøperne' [The Smell in Shops Can Calm Shoppers]. Internet article, www.forskning.no, 24 June.

Stålbom, Göran, and Birgitta Johansson, eds (2003). *Människan inomhus. Perspektiv på vår tids inneliv. En antologi* [The Human Indoors: A Perspective on the Interior Life of Our Age]. Stockholm: Formas.

Sundt, Eilert (1887 [1869]). *Om renligheds-stellet i Norge. Til oplysning om flid og fremskridt i landet* [On Cleanliness in Norway]. Fagerstrand: Bibliothek for de tusen hjem.

Süskind, Patrick (2015). *Perfume: The Story of a Murderer.* Translated by John E. Woods. London: Penguin.

Thorsheim, Peter (2006). *Inventing Pollution: Coal, Smoke, and Culture in Britain since 1800*. Athens: Ohio University Press.

Toremalm, Nils-Gunnar (2020). *En doft av Linné* [A Scent of Linnaeus]. Malmö: Sällskapet Heimdall.

Trodd, Colin (1994). 'Culture, Class, City: The National Gallery, London and the Spaces of Education, 1822–57'. Essay in *Art Apart: Art Institutions and Ideology Across England and North America*, ed. Marcia R. Pointon. New York: Manchester University Press, pp. 33–50.

Tunstad, Erik (2007). 'Sexy lukt – eller frastøtende?' [Sexy Smell – or Disgusting?]. Internet article, www.forskning.no, 18 September.

Twain, Mark (2018 [1871]). 'About Smells', in *The Curious Republic of Gondour and Other Whimsical Sketches*. Project Gutenberg, www.gutenberg.org, accessed 23 September 2025.

Uddgren, Gustaf (1920). *Strindberg the Man.* Translated by Axel Johan Uppvall. Boston: The Four Seas Company.

Vallée, Gérard, and Lynn McDonald, eds (2006). *Florence Nightingale on Health in India*. Volume 9 of the Collected Works of Florence Nightingale. Waterloo: Wilfrid Laurier University Press.

Van Campen, Cretien (2014). *The Proust Effect: The Senses as Doorways to Lost Memories*. Translated by Julian Ross. Oxford: Oxford University Press.

Verbeek, Caro (2020). *Ruiken aan de tijd. De olfactorische dimensie van het futurisme (1909–1942)* [Smelling Time: The Olfactory Dimension of Futurism]. Vrije Universiteit Amsterdam.

— (2021). 'In Search of Lost Scents: The Olfactory Dimension of Futurism'. Essay in *International Yearbook of Futurism Studies*, vol. x, ed. Günter Berghaus. Berlin and Boston: de Gruyter, pp. 247–75.

Vindheim, Jan Bojer (2005). *Historiens hemmeligheter. Myte, magi og mysterier fra Atlantis til Harry Potter* [History's Secrets: Myth,

Magic and Mystery from Atlantis to Harry Potter]. Oslo: Kagge forlag.
Vogt, Johan (1945). *33 Regler for god tone på Grini* [33 Rules for Good Form at Grini]. Oslo: H. Aschehoug & Co.
Vogt, Yngve (2019). 'Reformasjonen tok flere hundre år' [The Reformation Took Several Hundred Years]. Article in *Apollon*, 29/2 (April), at www.uio.no/forskning.
Vonnegut, Kurt (1969). *Slaughterhouse-Five.* New York: Dell.
Walsø, Margit (2022). *Din femte sans. Hvordan luktesansen beskriver deg, din fortid og din fremtid* [Your Fifth Sense: How Your Sense of Smell Describes You, Your Past and Your Future]. Oslo: Kagge forlag.
Wilde, Oscar (2010 [1890]). *The Picture of Dorian Gray*. London: Penguin.
Williams, Megan N., and Coren Apicella (2018). 'Synthetic Copulin Does Not Affect Men's Sexual Behavior'. Article in *Adaptive Human Behavior and Physiology*, 4, pp. 121–37.
Wilson, F. P., ed. (1925). *The Plague Pamphlets of Thomas Dekker*. Oxford: Clarendon Press.
Winner, Langdon (1986). *The Whale and the Reactor*. Chicago: The University of Chicago Press.
Wold, A. E. (1998). 'The Hygiene Hypothesis Revised: Is the Rising Frequency of Allergy Due to Changes in the Intestinal Flora?' Article in *Allergy*, 53, pp. 20–25.
Wolke, Lars Ericson (2013). *Bombkrigets historia. Taktik och terror under 100 år av flygkrig* [The History of Bomb Warfare: Tactics and Terror Over a Century of Aerial Warfare]. Lund: Historiska Media.
Yokoyama, Issei, et al. (2020). 'Inhalation of Odors Containing DMHF Generated by the Maillard Reaction Affects Physiological Parameters in Rats'. Article in *Scientific Reports*, 10, art. 13931, at www.nature.com.
Zjadan, Serhij (2022 [2005]). *Anarchy in the UKR*. Translated into Norwegian by Dagfinn Foldøy. Oslo: Pax forlag.

ACKNOWLEDGEMENTS

The book is based on contemporaneous historical sources – mainly fiction, including comedies and satires that are particularly concerned with smell. I have only exceptionally resorted to modern historical literature because it often overlooks the fact that the world does not just smell different from how it used to; the status of our sense of smell was also entirely different in the past. Honourable exceptions include a group of sociologists, social anthropologists and cultural historians who first became interested in the cultural history of smell in the second half of the twentieth century; these include Constance Classen of Canada, Trygg Engen of Norway and Alain Corbin and Robert Muchembled of France.

Many people have contributed to the work on this book, not only through their professional expertise but by taking part in on-the-spot explorations and descriptions of smells. Particular thanks are due to Anders Eik Pilskog, Andrea Bergh Pettersen, Anna Fara Berge, Dag Roalkvam, Jan Gunnar Skjeldsøy, Marie Rosenberg, Nina Castracane Selvik, Simon Holmen, Sofia Lersol Lund, Stian Tveiten, Svanhild Naterstad and Trond Berge.

Most of all, though, I would like to thank visual artist Anette Rosenberg for the illustrations she created for this book, which you will find here: https://reaktionbooks.co.uk/work/smell. They were developed as representations of colours, shapes and textures based on a painstaking and sometimes demanding exploration of smells.

Of course, it is something of a paradox to illustrate a book about smell in the first place. But it is an unfortunate fact that all engagement in our age must first go through the eye. My hope is that this will also stimulate the reader's sense of smell.